The Geometry of War

The Geometry of War
1500–1750

Catalogue of the Exhibition

Jim Bennett
Stephen Johnston

University of Oxford Old Ashmolean Building
Museum of the History of Science, Broad Street, Oxford

Published in the United Kingdom by the
Museum of the History of Science, Broad Street, Oxford.

First published in May, 1996.

A CIP catalogue record for this book is available from the British Library.

ISBN 0 903364 08 5

Thanks to:

Fédération Internationale des Géomètres
Brian Archer
The Operators, Oxford University Computing Services

Design: Giles Hudson
Photography: Jim Bennett & Lynn Norman
Editorial advice: Tony Simcock

Set in Monotype Janson and Times New Roman
Printed by Cheney & Sons Ltd, Banbury

An electronic edition of this catalogue may be found on the Internet at http://www.ox.ac.uk/departments/hooke/

Contents

Introduction

THE MATHEMATICIANS of the Renaissance applied their geometry to all manner of practical disciplines – from navigation and surveying to cartography, perspective and dialling. They aimed to demonstrate the usefulness of geometry as well as its ingenuity and certainty, and to associate it with action, achievement and progress. Many new instruments were designed in this context, as the collections of the Museum of the History of Science amply demonstrate.

Developments in the art of warfare in the late fifteenth and sixteenth centuries provided mathematicians with one such outlet for geometry. The ability to make a heavy gun in a single metal casting led to ordnance that was longer and capable of more accurate fire. Further, the addition of trunnions on which the barrel could turn in a vertical plane and the development of moveable carriages improved the setting of direction and elevation, and so encouraged the idea of predicting and calculating ranges. Practical mathematicians responded to these changes by devising techniques, designing instruments and writing books.

The speed with which the mathematicians responded to the novel set of problems thrown up by changes in the conduct of war is striking. It was in the sixteenth century that cannon came to be used in large numbers and their effect became critical to the outcome of military engagements; in the same period the geometry of war became a major branch of practical mathematics. To harness the capabilities of the new weaponry, gunners needed instruments to measure both the inclination of the barrel and the distance to the target, together with a means of relating these two measurements; the geometers offered a variety of solutions, as well as designs for fortifications to withstand attack from the new artillery. The frequent wars in sixteenth-century Europe added urgency to their work and helped justify claims for the importance of their discipline. As in other areas of practice, such as cartography or surveying, the mathematicians were quick to recognize an opportunity and they responded with enthusiasm – perhaps with over-enthusiasm, in view of the instruments and images on view in this exhibition.

The ingenuity and precision of the instruments on display, and the elegance, poise and delicacy of many of them, contrast with the harsh and uncompromising conditions of the battlefield. A review of such instruments is bound to raise the question of how useful they could really have been in practice. They were supposed to be employed in battle, but it is clear that their purported military value also had other functions – in justifying textbook geometrical problems, for example, and in attracting patronage.

The exhibition at the Museum of the History of Science seeks to draw attention to the fact that the advent, not of gunpowder directly, but of its more effective use in improved ordnance, precipitated a new variety of practical mathematics and a new disciplinary discourse that flourished through the sixteenth and seventeenth centuries. Though few historians of science or of art have given it much attention, this geometrical discipline engaged a great many practitioners in the period and was associated with a range of technological developments applied throughout Europe, both in military campaigns and in large works of military engineering. The exhibition itself has three main sections, dealing with gunnery, range-finding and surveying, and fortification. The mathematics of troop formations is the first of two additional topics, while the other is a reminder that the telescope – one of the most emblematic instruments of science – was originally introduced as an instrument of war.

Fig. 1 (previous page) A night scene of instrument use in the heat of battle. From catalogue no. 13.

Fig. 2 (above) Gunners measuring up a gun and its ladle. From catalogue no. 13.

Gunnery

The art of gunnery was complex and dangerous and the gunner's ability to fire reliably and accurately was frequently criticized. Although mathematicians could not remedy variations in powder or in the form of individual guns, they did seek to improve gunnery by devising instruments for the measurement of shot, the elevation of guns and mortars, and the calculation of the range of fire. Instruments for these operations include calipers, gauges, quadrants, sights, levels and specialized rules. Two or more of these elements were often combined in a single design, in a marketing initiative typical of mathematical instrument makers. Other mathematical instruments such as sundials were also incorporated in the more exotic hybrids, which typically date from the sixteenth or earlier seventeenth centuries. By the eighteenth century, as mathematical instrument makers became regular suppliers to professionalized ordnance departments, more standard patterns were emerging.

Early modern gunners had to cope with a bewildering variety of weaponry. Guns were named according to their size, weight and length, or by the type and weight of their shot. Each class of gun was different, and was primed with specific amounts of powder (often inserted with individualized ladles) before being loaded with its own shot. Printed sources frequently tabulated the appropriate information for each type of gun and some instruments likewise carry tables to serve as *aides-mémoire.* Instruments were also enlisted to match the correct size of shot to different guns. Calipers provided measurements of the diameters of gun bores and of spherical shot, and often directly indicated the weight of the shot. Gauging rods also accomplished the same linking of diameter and weight, and scales adapted for iron, lead and stone shot frequently appear on German instruments.

Although heavy guns were often fired for maximum destructive effect at point blank range (with the barrel horizontal), greater range could be achieved by elevating the gun. Quadrants, sights and levels enabled the gunner to set the barrel at specified elevations. These instruments were constructed in an enormous variety of forms and styles, with their scales diversely graduated in degrees, inches or 'gunner's points'. In his text of 1537 inaugurating the 'new science' of artillery, Niccolò Tartaglia described a quadrant which was inserted into the muzzle of the gun. Subsequent authors noted that this exposed the gunner to enemy fire and they offered alternative instruments which could be set up at the gun's breech. The many surviving styles of levels and sights fall into this safer category of instrument. Levels measure the inclination of the barrel using a plumb line or a rigid plummet and a graduated arc. Sights have a moveable pinhole or a series of fixed holes through which the gunner observed, making an alignment with the top of the muzzle.

Fig. 3
Elevating a mortar by lamplight for a bombardment at night.
From catalogue no. 13.

To determine appropriate elevations for his cannons and mortars, a gunner had to know the distance of the target and also be able to relate this range to the elevation of the piece. The question of determining the correct elevation necessary to fire a shot a given distance (and its inverse, the prediction of range at a given elevation) was the most taxing problem of gunnery as a mathematical art. Gunners had their own rules of thumb, and instruments were used to embody rules relating range and elevation. But questions of ballistics also engaged the most prominent mathematicians. Tartaglia set the terms of the debate by seeking to portray the geometry of a projectile's trajectory based on the opposed natural and violent motions of Aristotelian physics. His work provided the basis for many subsequent accounts in textbooks and manuals. Galileo offered a new foundation in his *Discorsi* of 1638, demonstrating the parabolic path of projectiles and reinforcing the military relevance of his work with a complete table of ranges. Through the seventeenth and eighteenth centuries mathematicians of the highest stature, such as Newton, sought more accurate depictions of projectile trajectories by attempting to take into account disturbing factors such as air resistance. Their sophisticated efforts may have eluded the grasp of the average gunner, but do demonstrate that the study of projectile motion was a point of intersection for the art of war, the mathematical sciences and contemporary natural philosophy.

Range-finding and Surveying

How was the gunner to determine the distance of his target? The traditional method of linear measurement in land surveying was simply to lay ropes or poles between the two stations concerned – hardly an option when the distant station was a hostile position. However, sixteenth-century geometers were seeking to introduce the technique of triangulation, and range-finding was part of their case for a new geometry of surveying. Distant stations could be located by sighting from either end of a measured baseline; their distances were found by measuring the angles formed with the baseline and by subsequent calculation, or by a more straightforward graphical method. In ordinary surveying, the possibility of triangulating as many features as were required from a single linear measurement was presented as an innovation that would greatly improve efficiency and convenience. Range-finding offered a particularly appropriate application, since access to the target feature was not simply inconvenient, but impossible. Further it gave the designers of new surveying instruments an immediate and telling example of the value of the novel method, so that relevant instruments – particularly the universal instruments, with all manner of claimed uses – were frequently illustrated in dramatic and urgent action.

Two types of surveying instrument were applied to triangulation and illustrated in use in warfare. In the first group, comprising what are called 'triangulation instruments', parts of the instrument are arranged to form a triangle similar to that on the ground, for example by aligning rules with the lines of sight from the baseline. Scaled measurements can then be taken directly from graduations on these arms. The other category of instrument employs scales typically found on the backs of astrolabes – the circular degree scale and the shadow square or geometrical quadrant. This class includes, of course, the astrolabe itself, but extends to the theodolite, circumferentor and graphometer.

Such instruments were made relevant to military life not only through the range-finding of the gunners. It was claimed that they were generally helpful to the military surveyor, either when laying out a new fortification, or when measuring and representing an existing one. Accounts of the instruments show them being used in such contexts and point to the particular advantage of being able to survey from a safe distance.

Fortification

A further relevant aspect of military surveying is the design of fortifications. If guns precipitated new branches of the mathematical sciences dealing with projectile motion and with range-finding, they also created the conditions for a new military architecture. Just as contemporary civil architecture was founded on geometry, expressed through the classical style, so too the new genre of fortification rested on a geometrical formalism.

The high walls of the medieval fortress were good for repelling attack from beneath, but were vulnerable to heavy guns: they presented large targets without providing suitable platforms for defensive fire. Yet if walls were to be low and stout, so as to withstand artillery bombardment, how were they to be defended against direct infantry assault? The new style of fortification emerged as a response to this problem. The solution was to create squat, thick walls that were defended by sidelong or flanking fire aimed from projecting gun emplacements or bastions. These bastions had to offer protection to the adjoining walls on either side and the whole fort had to be enclosed, resulting in polygonal outline plans. The angled shape of the bastions was progressively refined so that each bastion offered covering fire to its neighbours and no 'dead space' remained hidden from defensive fire. Beyond the primary wall and bastions, further positions and defensive structures could be extended outwards. These limits and rules set the conditions for the development of elaborate geometrical designs, not always constrained by the features of actual sites and the limited budgets of genuine commissions.

Such celebrated practical geometers of the Renaissance as Francesco di Giorgio, Filippo Brunelleschi, Leonardo da Vinci, Albrecht Dürer and Simon Stevin were concerned with the solutions to these problems, as were artists who might not be immediately associated with a programme of this kind, such as Bramante and Michelangelo. A species of practical geometry that began in Italy in the later fifteenth century, spread to other parts of Europe in the sixteenth, supported by the design of instruments and the publication of books. The general recommendation was for arrow-headed bastions projecting from the corners of a walled polygon, but beyond this there was plenty of scope for individual styles and schools. In

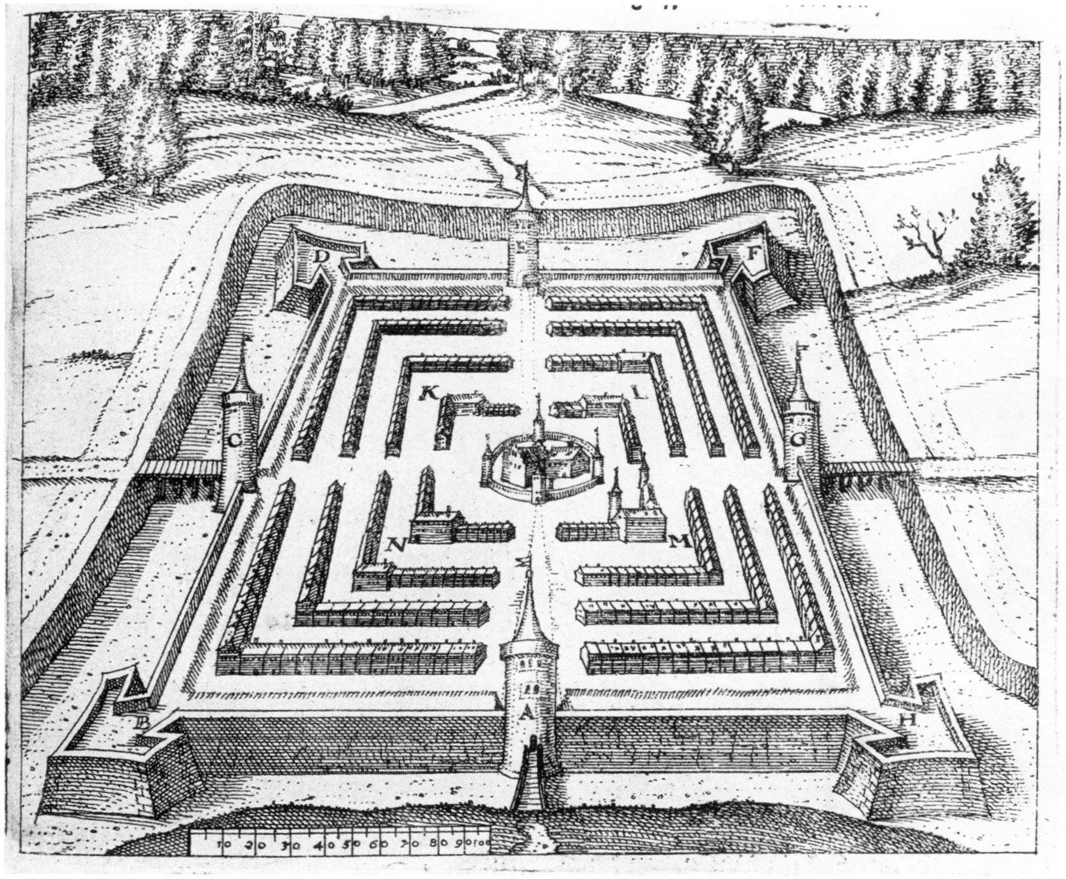

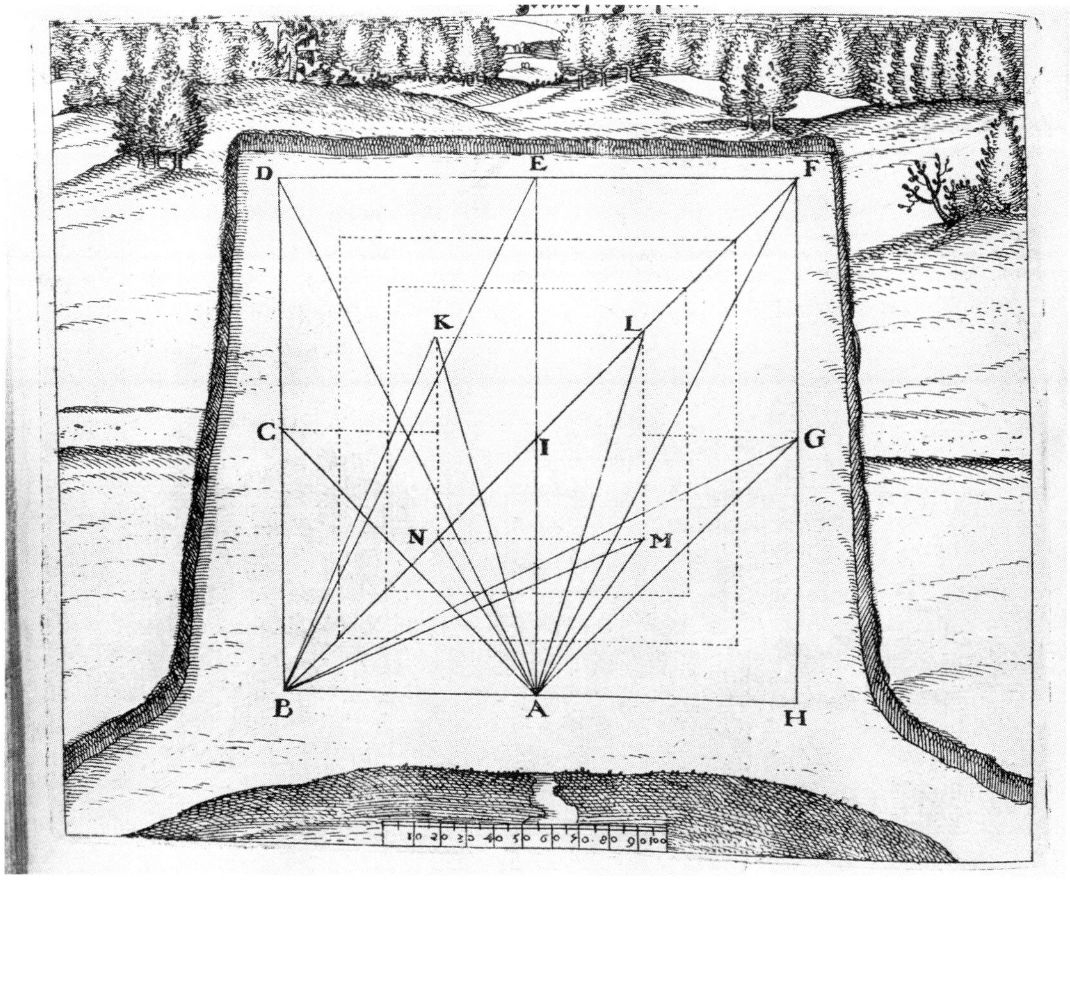

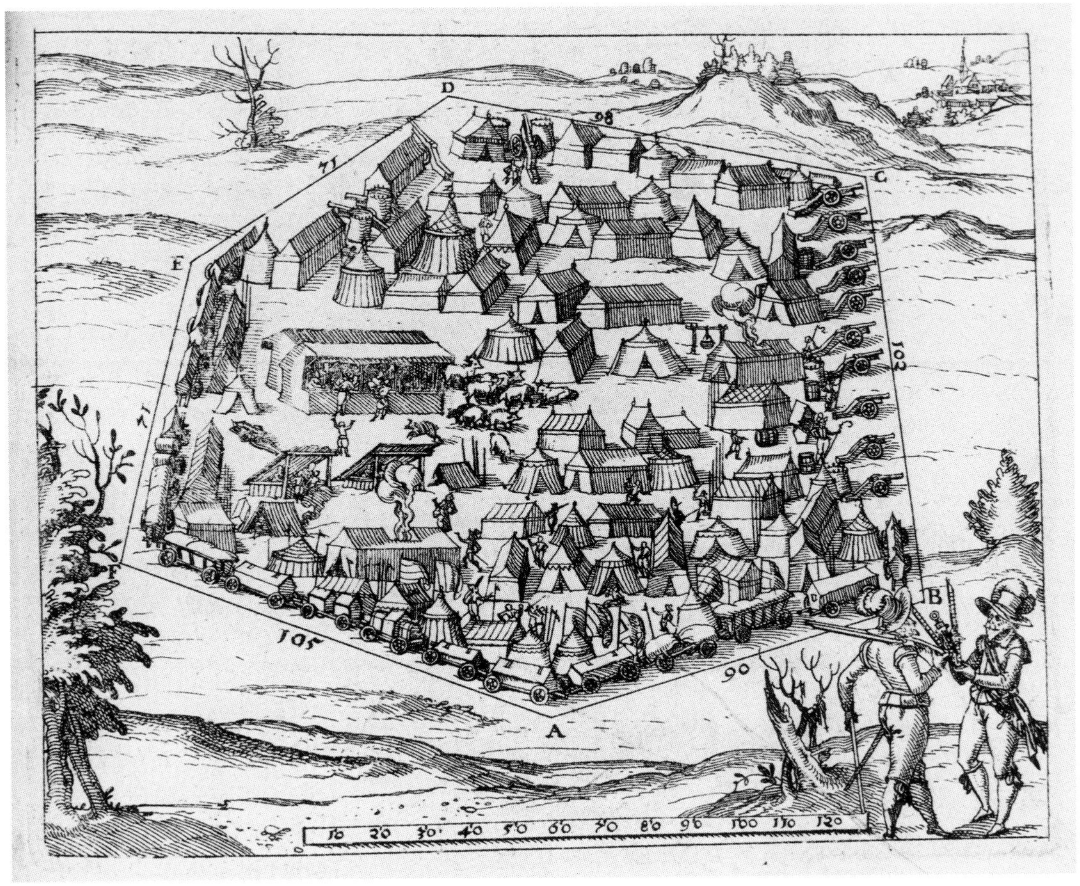

Fig. 4 (opposite top)
A fortified town with angled bastions.
From catalogue no. 58.

Fig. 5 (opposite bottom)
A survey of the geometrical plan for the town in figure 4.
From catalogue no. 58.

Fig. 6 (above)
Military encampments as well as fortifications were to be laid out on polygonal ground plans with the help of geometrical instruments.
From catalogue no. 13.

practice the basic polygon often had to accommodate the natural conditions of the site or its existing structures. The French school took the lead in the development of such systems in the second half of the seventeenth century, with the practice and theory of Sébastien le Prestre de Vauban dominant into the eighteenth. The outcome was a vast programme of work and a thriving specialist geometrical discipline.

Geometers had to consider not only the plan of an ideal fortification: the section offered them further possibilities. Fortress defenders wanted both to impede the progress of any attackers and to oblige them to present an advantageous target for their guns. A large ditch in front of the rampart was fundamental to defence, and the earth from the ditch provided an embankment on the outer border that both obscured the walls and provided a sloping approach, angled so as to be swept by defensive fire. Other features could be incorporated, such as the covered way on the outer side of the ditch, for patrols shielded from fire. As with the plan or 'trace', the result was an elaborate structure subject to revision and adjustment by different practitioners.

A Practical or a Polite Science?

As a new branch of mathematical science, the geometry of war was ambiguously positioned in a number of respects and these ambiguities deserve attention from historians. Was this a practical science or a polite one? Did it belong at war or at court? Was it driven by practice or by theory? Was it characterized by action or by rhetoric? Was it part of mathematics or of natural philosophy? Answers to each of these questions fall somewhere between the two respective alternatives, and their positions on each of these ranges of possibilities change over time. It is within these ambiguities and changes that some of the most interesting aspects of the subject are to be found.

The tension between the practical and the polite is perhaps most obvious in the contemporary literature. Instruments are illustrated in use where conflict is imminent or already begun, and the coolness of the practitioners who apply their geometry in the heat of battle can seem improbable. Such examples of purported use initially served to stress the potential importance of the practitioners' geometrical techniques. With the acceptance that aspects of warfare were amenable to geometry, the association could be used to justify a group of mathematical problems that in practice would remain confined, for the great majority of readers, to textbook instruction.

The same can be observed in the progress of popular interest in fortification. Both gunnery and fortification become established in general encyclopaedic works aimed at the gentle market, and polygonal scales are included in a conventional way on a variety of instruments whose links to professional military surveying seem implausible. By the end of the period covered by the exhibition, popular lecturers in eighteenth-century London, such as Benjamin Martin and Erasmus King, were offering fortification as mathematical entertainment. Another lecturer, W. Griffiss, tackled the subject in a tavern in the evening, 'illustrated by various Plans, and a large Model of a fortified Town', and repeated the performance the following morning 'for the better Accommodation of the Ladies' (Morton & Wess, p. 73).

Military mathematics likewise featured in academic as well as coffee house teaching. To take a local example, David Gregory proposed a new mathematics course for Oxford in 1700 which included 'a lecture on fortification, so far as 'tis necessary for understanding it without actually serving in an army or forttifying a town or camp' (quoted in Bryden). The museum building itself witnessed instruction in fortification in the very room where the present exhibition has been mounted, when John Whiteside introduced the subject into his course on mathematics in 1723. At the same time as these popular and academic activities, there was at the practical level an active programme of fortress construction across Europe, and this lent credibility and status to the polite interest.

It is probably the surviving instruments that most eloquently address the tension between the courtly context and the battlefield. Elegant and ingenious instruments, perhaps in gilt brass and accompanied by tooled leather cases, were far beyond the material and probably the educational resources of the ordinary gunner. They could, however, enhance the image of an active officer abreast of the finer points of contemporary warfare. The maker offered a rare accessory to enhance a courtly posture and in exchange the patron offered a valuable and perhaps prominent commission. Instrument design might itself accommodate a rhetorical or gestural aspect, as when mathematical instruments are made to resemble, or even to act as, weapons. Yet not all surviving instruments fall into this class: they range from the improbably ornate to the crude and humble, and we must remember that the former are much more likely to have found their way into museum collections. We know, for example, that the British Board of Ordnance ordered gunner's quadrants and perpendiculars in quantity. On the other hand again, John Muller, Professor of Fortification and Artillery at the Royal Military Academy, Woolwich, says that gunners should learn to judge by eye and dismisses the use of instruments in his *Treatise of Artillery* of 1757.

The geometry of war sits ambiguously too between practice and theory. It is an ambiguity characteristic of the mathematical sciences and one from which they derived their potential for influential development in the Renaissance. Leonard Digges points clearly to this in his *Pantometria* of 1571, where he asserts that the perfection of gunnery is beyond both the mathematical novice, 'though hee turmoile in powder and shot all the dayes of his life', and equally the geometer, 'leaning onely to discourse of reason', who 'shall fall into manifolde errors, or inextricable Laberinthes.'

Fig. 7
Checking that a gun's bore is true; instruments used together with a tool inserted into the muzzle.
From catalogue no. 13.

The tension between deeds and words, between action and rhetoric, is particularly evident in contributions from the practitioners themselves, or at least from writers who present themselves as practitioners. Robert Norton, who describes himself as 'one of his Maiesties Gunners and Enginiers' on the title-page of *The Gunner* of 1628, asserts firmly that '… lead on by Experience the Mistris of all Arts, Action being the best Tutor … I … haue endeuoured herein more to respect a few experimented truthes, then many Rhethoricall imbellishments of words.' An 'experiment' at the time has the sense of a deed or venture, as distinct from a discussion or projection, without the more specific reference it came to have in later science. But it is clear that the sentiment of the Royal Society's motto, 'nullius in verba', had an earlier currency in this empirical tradition.

Finally there is the changing position of gunnery between mathematical science and natural philosophy. The former of these separate branches of Renaissance learning dealt with mathematics and its practical applications, while the latter was concerned with causal explanations of natural phenomena. Norton again was bullish and provocative: gunnery, he said, was a profound study, 'euen able to spose the knowne parts of Naturall Philosophy, Arithmetick, Geometry, and Perspectiue, each of which her handmayd is'. In reality the relationship was much more ambiguous,

Fig. 8
Gunners at work measuring a barrel, with matching measured ladle in the foreground.
From catalogue no. 13.

Fig. 9
Aiming at a target with the help of a sight mounted at the breech end of a gun.
From catalogue no. 13.

Fig. 10
Geometrical instruments placed centre stage amongst the implements of war.
From catalogue no. 13.

but the fact that this aspect of practical geometry raised questions pertinent to natural philosophy may be relevant to the emergence of a new relationship between the two – one where mathematics, experiment and instruments are all recruited to a reformed explanatory programme for the natural world.

The exhibition concludes with one such connection. Galileo began his career as a mathematical scientist, a teacher of practical mathematics, including fortification, and a designer of mathematical instruments. His ambitions, however, were towards the higher science of natural philosophy and one instrument he was involved with – the telescope – though it began as an instrument of warfare, became in his hands a reforming tool of natural philosophy. Reform was foreshadowed in two ways – through the telescopic observations Galileo marshalled in support of Copernicus, and through his use of an instrument as a means of investigating the natural world.

The telescope was a new departure in the methodology of natural philosophy, but it was through projectile motion that the discipline was most directly implicated in the geometry of war. Early writers on ballistics – Tartaglia and his successors – couch their discussions in terms of the natural and violent motions of Aristotelian physics. The form of the ballistic trajectory, traced by an initially rising projectile which then begins to falter and to descend, was presented as the combined outcome of the two opposed motions and the transition between them. The superiority of Galileo's technical achievement in the *Discorsi* should not blind us to the significance of a century of problem setting and the significance too of setting the terms of a solution – that is, as a geometrical trajectory with universal reference yielding predictable consequences. The treatment from Tartaglia onwards is thus characteristic of the mathematical sciences and particularly of their most sophisticated branch, astronomy.

Although Newton's success in *Principia* is often presented as a transcendent achievement of the geometry of reason, part of the prior development that made 'the mathematical principles of natural philosophy' a viable conjunction was contained in the geometry of war. Almost two hundred years after the Renaissance discourse of ballistics was opened by Tartaglia, Newton, in his *System of the World* (published posthumously in 1728), explained the dynamics of planetary motion in terms of a body projected horizontally from a mountain top with successively greater initial velocity. The problem is the old one of point-blank range but, not only has Newton escaped from the language of gunnery into that of abstract mechanics, the range of his projectile extends until, instead of descending to earth, it falls forever in an orbital trajectory.

Catalogue

1. Niccolò Tartaglia, *La nova scientia* (first edition, 1537; this edition, Venice, 1558). Bound with his *Quesiti et inventioni diverse* (first edition, 1546; this edition, Venice, 1554).

Few European mathematicians of the 16th century had been as directly affected by war as Niccolò Tartaglia. In 1512, when still a boy, he received a facial wound during the sack of Brescia by the French. Left with a speech defect he adopted the nickname of Tartaglia ('stammerer') to replace his original surname Fontana.

If Tartaglia's very identity was marked by war, he in turn reshaped the character of military discourse by identifying a 'new science' of artillery and casting it as a mathematical discipline. As a mathematician he was first directed to military questions in 1531 or 1532 in Verona when he was consulted on the maximum range of cannon. Following up this initial query he subsequently published what was proudly announced as his *Nova scientia* of 1537. Tartaglia offered a general account of the fall of heavy bodies and a specific consideration of the path of artillery shot. His novel mix of natural philosophy and mathematics heralded the wider significance that falling bodies would soon have in European intellectual culture through the work of Galileo. But Tartaglia also showed how to measure the elevation of an artillery piece using a gunner's quadrant, and provided a discussion of surveying for range-finding.

After *La nova scientia*, Tartaglia returned to military matters in the *Quesiti et inventioni diverse* (1546), further discussing topics such as the shape of a projectile's trajectory. Although he qualified the theoretical accuracy of his depiction, for practical purposes Tartaglia consistently described the path of the shot in three parts: an initial straight line propelled by the force of the gunpowder; a curving section as gravity began to work its effect; and finally a straight line again as the shot fell perpendicularly to the ground. A measure of Tartaglia's importance for the study of artillery is that this account was still being paraphrased and parroted into the later 17th century. But he did not provide later authors with all they could have wished. In both the *Nova scientia* and the *Quesiti et inventioni diverse* he promised tables of 'randons' (ranges achieved at different elevations), but these were never forthcoming.

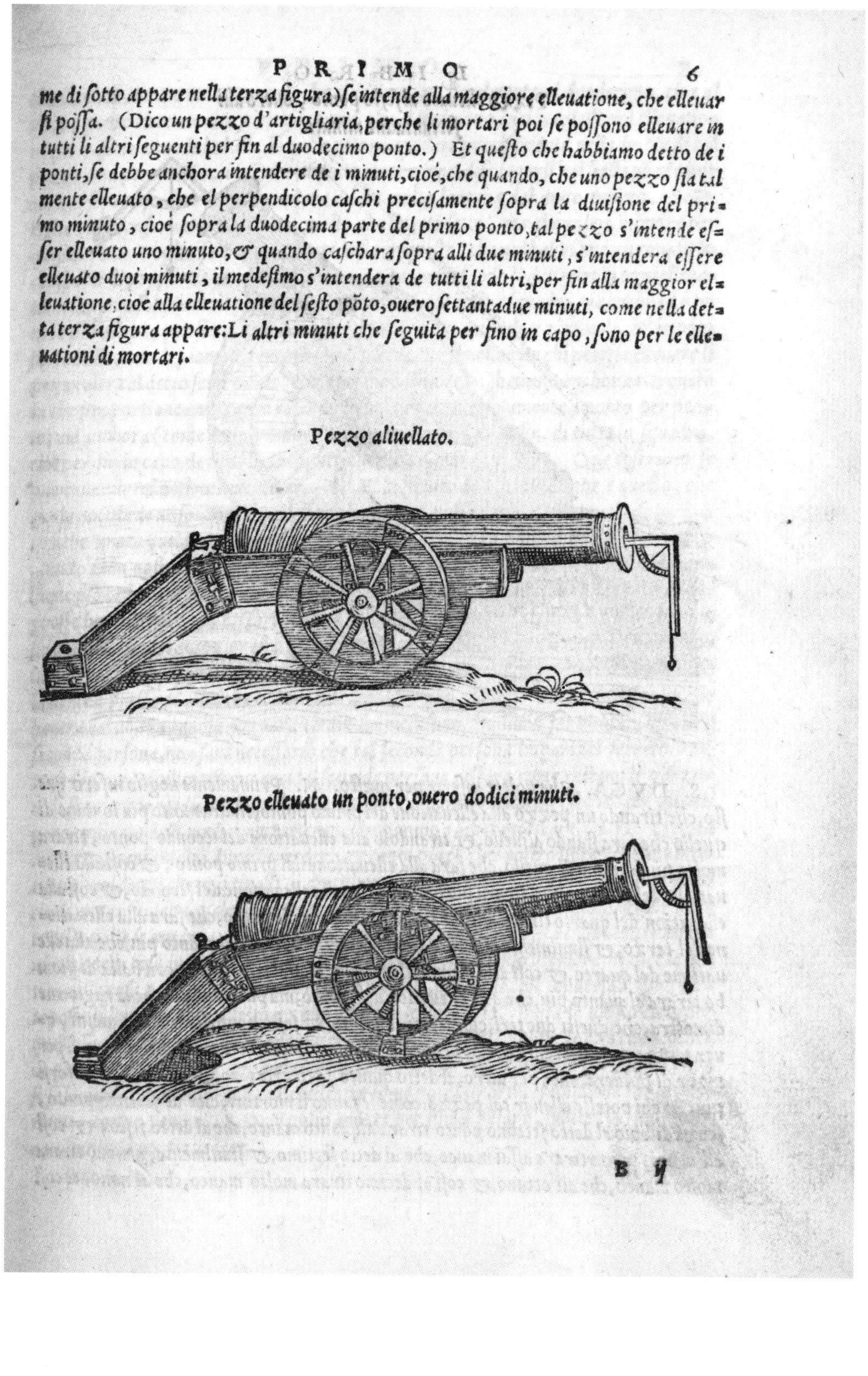
PRIMO 6

me di ſotto appare nella terza figura) ſe intende alla maggiore elleuatione, che elleuar ſi poſſa. (Dico un pezzo d'artigliaria, perche li mortari poi ſe poſſono elleuare in tutti li altri ſeguenti per fin al duodecimo ponto.) Et queſto che habbiamo detto de i ponti, ſe debbe anchora intendere de i minuti, cioè, che quando, che uno pezzo ſia talmente elleuato, che el perpendicolo caſchi preciſamente ſopra la diuiſione del primo minuto, cioè ſopra la duodecima parte del primo ponto, tal pezzo s'intende eſſer elleuato uno minuto, & quando caſchara ſopra alli due minuti, s'intendera eſſere elleuato duoi minuti, il medeſimo s'intendera de tutti li altri, per fin alla maggior elleuatione, cioè alla elleuatione del ſeſto põto, ouero ſettantadue minuti, come nella detta terza figura appare: Li altri minuti che ſeguita per fino in capo, ſono per le elleuationi di mortari.

Pezzo aliuellato.

Pezzo elleuato un ponto, ouero dodici minuti.

B ij

QVESITI ET INVEN-
TIONI DIVERSE
DE NICOLO TARTAGLIA,
DI NOVO RESTAMPATI CON VNA
GIONTA AL SESTO LIBRO, NELLA
quale si mostra duoi modi di redur una Città inespugnabile.
LA DIVISIONE ET CONTINENTIA DI TVTTA
l'opra nel seguente foglio si trouara notata.
CON PRIVILEGIO
APPRESSO DE L'AVTTORE
M D L I I I I.

Fig. 11 (left)
The use of the gunner's quadrant explained.
From catalogue no. 1.

Fig. 12 (right)
Niccolò Tartaglia, the 'stammerer'.
From catalogue no. 1.

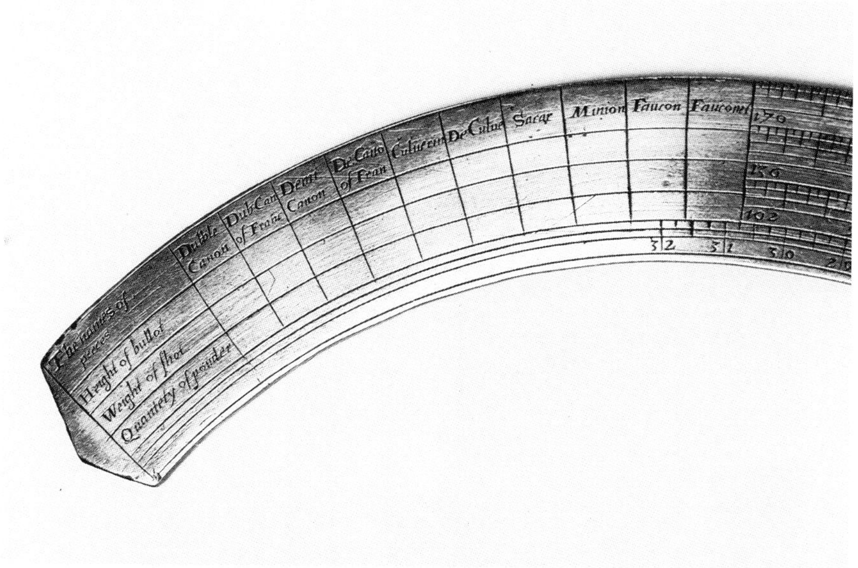

2. Sector

c. 1600

Large guns of the Renaissance came in a bewildering variety of forms. They were variously categorized by their length, their weight and the size or weight of the shot they fired. Each gun was unique because, after it had been cast from bronze or iron, its mould had to be smashed to extract the piece. But although standardization was not possible, different types of 'great ordnance' were recognized and named. To keep track of the interrelations between powder, shot and piece, gunners and administrators compiled tables listing the ammunition and supplies appropriate to each gun. Such *aides-mémoire* became standard features of manuscript and printed works, but instruments also regularly served as a convenient and robust location for them.

The arc of this early English sector carries an artillery table drawn up to record the size and weight of shot and the amount of powder required for different types of artillery. The table's data were never entered but 'The names of peeces' were listed in order of decreasing size: double cannon, double cannon of France, demi-cannon, demi-cannon of France, culverin, demi-culverin, saker, minion, falcon and falconet.

The sector was a new instrument of the late 16th century, first publicized in a book by Thomas Hood published in London in 1598. Equipped with sights for use by surveyors it was also provided with various engraved scales for calculation and measurement. This unsigned English example is closely related to the form described by Hood and includes sectoral scales for the graphical calculation of proportions and for drawing polygons. The reverse of its arc carries a scale of degrees, subdivided by transversals for greater precision, along which is an additional series of points for setting out polygons.

Leg length (closed): 279 mm
Inventory no. 44,505

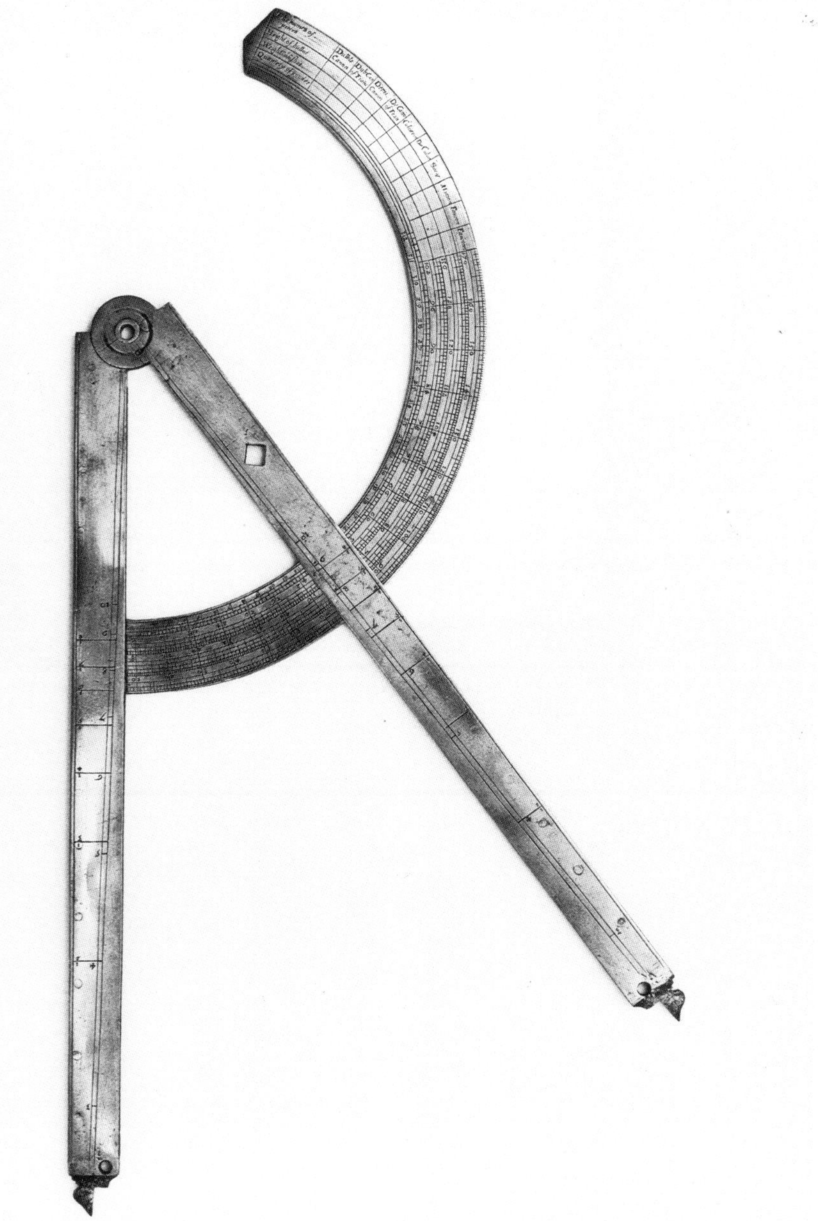

Fig. 13 (top)
Detail of incomplete gunner's table.
Catalogue no. 2.

Fig. 14 (bottom)
English sector, *c.* 1600.
Catalogue no. 2.

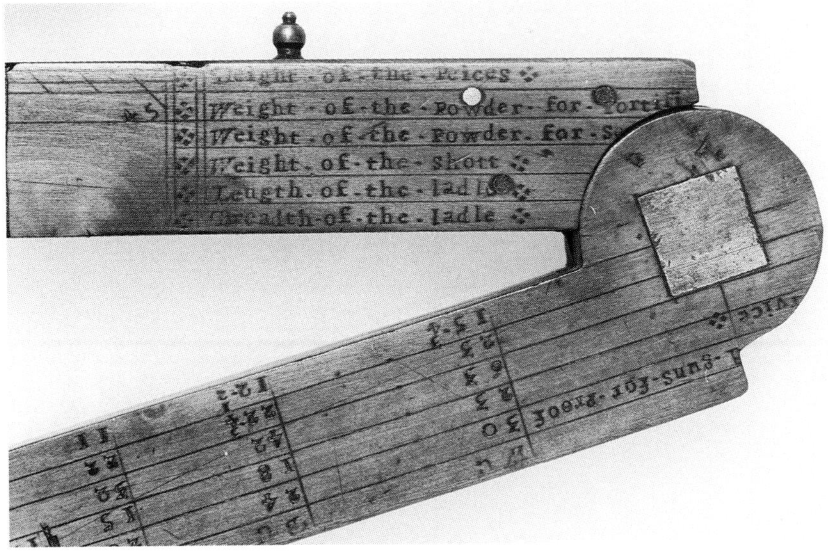

Fig. 15 (top)
Gunner's folding rule, 1680.
Catalogue no. 3.

Fig. 16 (above)
Headings from one of the artillery tables.
Catalogue no. 3.

3. Gunner's Folding Rule
Signed: Bengamin Jobson 1680

Wooden folding rules were familiar craft instruments in 17th-century England. Although most commonly meant for carpenters, the Jobson example suggests that gunners also profited from the flexibility of this format. The rule incorporates two artillery tables, the first of which includes abbreviations for many of the same named artillery pieces as appeared on catalogue no. 2. However, it also includes newer types of artillery rated only by the weight of their shot (3, 6, 8, 12, 16 and 24-pounders). Unlike the earlier English instrument, the artillery table on the Jobson rule is complete, with figures provided for the weight of shot and the dimensions of ladles appropriate to different artillery pieces. The weight of powder required for both ordinary use ('service') and the larger amount needed for preliminary testing ('proof') is also tabulated. Further help for the gunner is supplied on the reverse of the instrument with a table relating the size of iron cannon balls to their weight, ranging from 1 to 8 inches diameter.

In addition to its useful tables, the folding rule also carries a scale of inches, a double logarithmic line of numbers for calculation with dividers, and a quadrant scale. To use the instrument as a quadrant, the thin brass arm is set at 90° to the wooden legs. With a plumb-line hung from the tip of the arm, readings are taken against the degree scale on the supporting wooden arm.

Length (closed): 240 mm
Inventory no. 37,620

4. Gunner's Gauge

Stamped with a rosette
Dated: 1612

Artillery tables show that gunners typically matched gun and ammunition by the weight of the appropriate shot. Rather than directly weighing with a balance, a quick method of determining the weight of a ball was to measure its diameter with calipers and then to set this dimension against a specially constructed scale of weights. Scales giving the weight of shot of different materials are a particular feature of German artillery instruments. This is the plainest version, in which the scales are engraved on a rod of square cross section. Shot made of either stone (STEIN), iron (EISEN) or lead (BLEY) can be measured, while the fourth side carries a foot rule (SCHVCH).

Length: 306 mm
Inventory no. 53,100

5. Astronomical Compendium

Signed: Pragae fecit Erasmus Habermehl
c. 1600

Although contemporary with the gunner's gauging rod (catalogue no. 4), this compendium with string-gnomon sundial conjures up a world far removed from the scene of battle. Its plethora of tables and scales and its ingenious book-bound form would likely have appealed to a wealthy patron, enthused by its many operations and finely finished appearance. Yet it also carries on its longer bevelled edges iron, lead and stone scales as would be found on a gunner's gauge. Entered in Latin as *ferrum*, *plumbum* and *lapis* these scales are not, however, all they seem. In the short length available to them, they might be expected to provide only the first portion of the scales found on a longer instrument, thus serving only for small shot. However, Habermel has divided the scales differently, either graduating them in unusual units of weight or, more probably, shrinking the standard scales so that he can offer the same weight range in a constricted space. The penalty for this miniaturization is that the measured diameters would have to be reduced in the same proportion. Although this could be done with a pair of proportional dividers, the process is at best a clumsy one, suited more for demonstration than practice.

Dimensions: 106 x 60 x 12 mm
Billmeir Collection
Inventory no. 44,392

6. Protractor and Gunner's Gauge

Signed: I. G. Mettel fecit Nbg [Nuremberg]
18th century

A German semi-circular protractor whose arm carries gunner's scales for stone, iron and lead. The small size of the instrument limits the length of the gauge scales and means that only small shot can be measured. Unlike the Habermel compendium (catalogue no. 5) these scales are meant to give direct weights in ounces when a measured diameter is read against them.

Length: 129 mm
Inventory no. 45,186

Fig. 17 (above)
Gauge scale for lead on compendium by Erasmus Habermel.
Catalogue no. 5.

Catalogue no. 5 is also illustrated in colour on page 37 with catalogue nos 4 and 6.

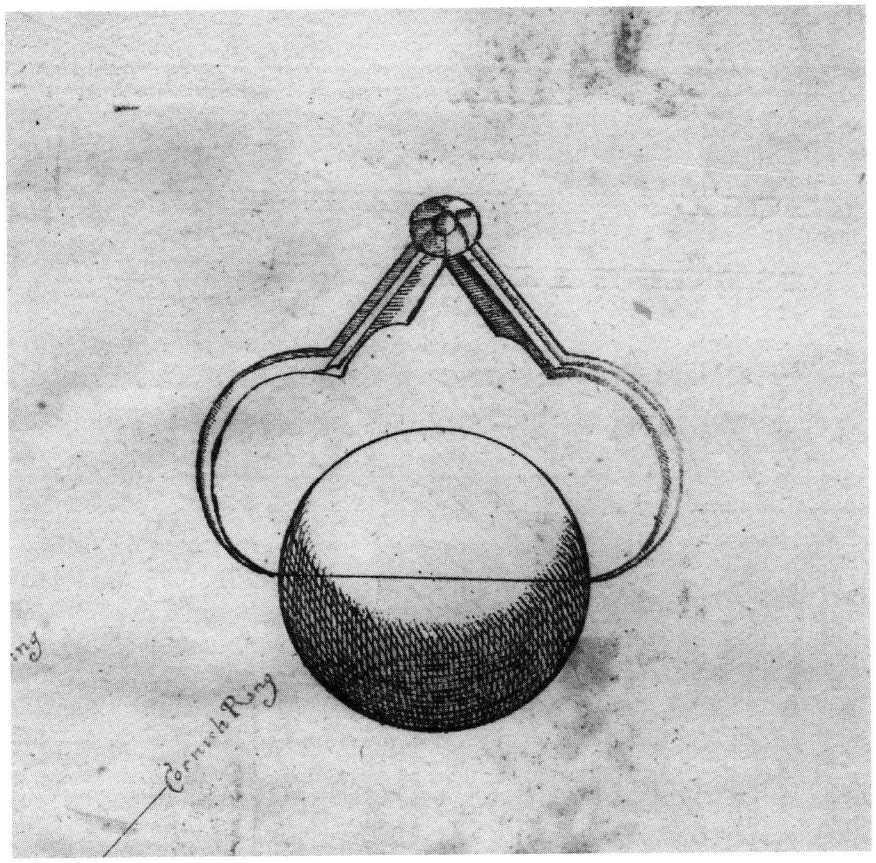

Fig. 18 (top left)
Taking the diameter of shot with gunner's calipers.
From catalogue no. 41.

Fig. 19 (top right)
Gunner's calipers.
Catalogue no. 8.

Catalogue no. 7 is illustrated in colour on page 40.

7. Gunnery and Dialling Instrument
Dated: 1595

When using a gunner's gauge, the diameter of the shot had to be taken with a pair of dividers and transferred to the gauge's scale in order to read off the weight. This German instrument represents an early attempt to provide a direct reading of weight simply from the separation of divider points.

The instrument consists of two main legs whose joint slides in the slot of a third, central leg. There are smaller link pieces connecting the side legs to the central leg. Alongside the central leg's slot are the three standard artillery scales for iron, lead and stone as well as a scale of inches. As the legs are opened or closed, the joint (which carries a compass box) moves in the slot and the edge of the compass box acts as an index to the four scales. Thus if the points on the main legs are set to the diameter of a cannon ball, the weight of the shot can be read directly from the scales.

Although the instrument is an ingenious combination of dividers and gunner's gauge, it has a drawback: since the main legs are straight, the points cannot themselves be used to measure round shot. However, this difficulty may not have mattered to the instrument's owner, as the device was presumably not a working tool: as well as its gunner's calipers, the instrument also incorporates a string-gnomon sundial and supplies tables of astronomical and calendrical information engraved on its main legs. Unusual though it is, there is at least one other surviving instrument of this pattern: a more finely executed version by Erasmus Habermel, which provides different tables (Eckhardt, p. 58).

Length: 186 mm
Inventory no. 52,589

8. Gunner's Calipers
c. 1700?

As ordinary dividers cannot readily measure round shot, specially shaped calipers were typically used (see figure 18). However some alternative methods were also tried. Rather than taking dimensions with points at the end of curved or straight legs, this English instrument clasps a shot between its two brass arcs. The arcs are hinged to pass freely over each other so that, when clasped around a ball, the inside edge of each arc intersects a graduated scale on the other. On one side are abbreviations for named artillery pieces, from falconet (FN) to whole cannon (WC, where the ends of the arcs meet). The other side is for guns rated by the weight of their shot, from 3 to 24-pounders.

Length: 260 mm
Inventory no. 42,060

9. Gunner's Calipers

Signed: N. Bion. ÆParis

c. 1700

To accommodate irregularities, and for the sake of safety, the shot of early modern artillery did not fit tightly in the bore of a gun. The difference between the diameter of the shot and the calibre of the gun was known as the windage. Using the hinged crosspiece and a single series of calibrated notches, these calipers by Nicolas Bion can measure both shot and calibre. Diameters taken between the inner ends of the instrument's points provide the weights of shot from ¼ to 48 pounds. At any given setting for the diameter of shot, the outer tips of the points give a slightly larger dimension, indicating the bore of the corresponding artillery piece; the same scale then indicates different guns from ¼ to 48 pounders, and also allows for windage. This dual measuring function is indicated by the inscriptions on either side of the instrument: 'diamettre des poids des bouleits' and 'Calibres des pieces'.

Bion illustrated this type of gunner's calipers in his *Traité de la construction et des principaux usages des instrumens de mathematique* of 1709 (catalogue no. 32; figure 46).

Length (closed): 163 mm

Inventory no. 43,257

10. Gunner's Calipers

Signed: Watkins & Smith London

c. 1770

Although gunner's calipers of the continental pattern were known in England through Stone's edition of Bion's book on mathematical instruments (catalogue no. 32), English-made calipers had already settled on a distinctive design. Their two flat blades are hinged so that they slide freely over each other, permitting the steel tips to point either inwards to take the diameters of shot or outwards to measure the bore of a gun. In this example, scales at the hinge give diameters in inches and the corresponding weight of iron shot. The scales for inches of the bore and for the pounder rating of the gun are engraved along the length of one of the limbs. The remaining space on the limbs is occupied with tables of specific weights and quantities of powder for iron and brass guns, as well as a sectoral scale of lines and a series of diagrams representing proportional relationships between different planes and solids.

Length (closed): 175 mm

Inventory no. 41,742

Fig. 20 (opposite left)
Gunner's calipers by
Nicolas Bion.
Catalogue no. 9.

Figs 21 & 22 (right)
Gunner's calipers by
Watkins and Smith.
Catalogue no. 10.

Figs 23 & 24
Gunner's sector by
Tobias Volckmer.
Catalogue no. 11.

11. Gunner's Sector

Signed: Tobias Volckmer. Sereniss: Ducis Bauarorum.
&c: Mathe: & aurifaber faciebat. .1618.

The early English versions of the sector were oriented towards use in surveying and found military relevance chiefly through their sectoral or arc scales for polygons. Only the occasional addition of tables for gunnery suggested a more explicit warlike purpose (catalogue no. 2). However, an alternative design of the sector by Galileo was independently devised for more directly military needs. Galileo's sector was, in its early version of 1597, created to provide a combined quadrant and calculating instrument, able to take gunnery elevations and compute the weights and volumes of shot. He subsequently developed his sector into a more general-purpose device for computation, with gunnery just one element of its calculating repertoire. It was in this latter form that the instrument was published in an Italian edition of 1606. Wider European recognition began with a Latin plagiarism of 1607 and continued with a full Latin translation in 1612.

German authors and instrument makers were quick to adopt and adapt the Galilean sector. This instrument shows that by 1618 sectors quite distinct from the Galilean pattern in both construction and scales were being produced. The instrument's pairs of sectoral scales provide for volumetric reckoning and also the construction of polygons. In addition there are gunner's gauge scales for the familiar materials of iron, lead and stone, each graduated in Munich units. Amongst the instrument's other features is a quadrant scale graduated to give the angle between the two legs and a compass surrounded by scales of degrees and polygons. Each leg also carries a scale marked 'PVNCTA DIVISA. IN ARTHOLORIA. VTENDA.' which enables the instrument's points to be used as direct measuring calipers.

Length (closed): 356 mm
Inventory no. 38,366

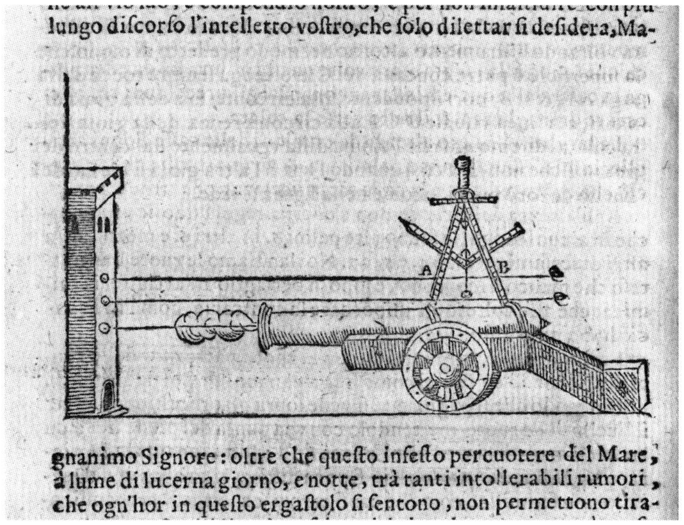

Figs 25 & 26
Bartolomeo Romano's
'proteo militare'.
From catalogue no. 12.

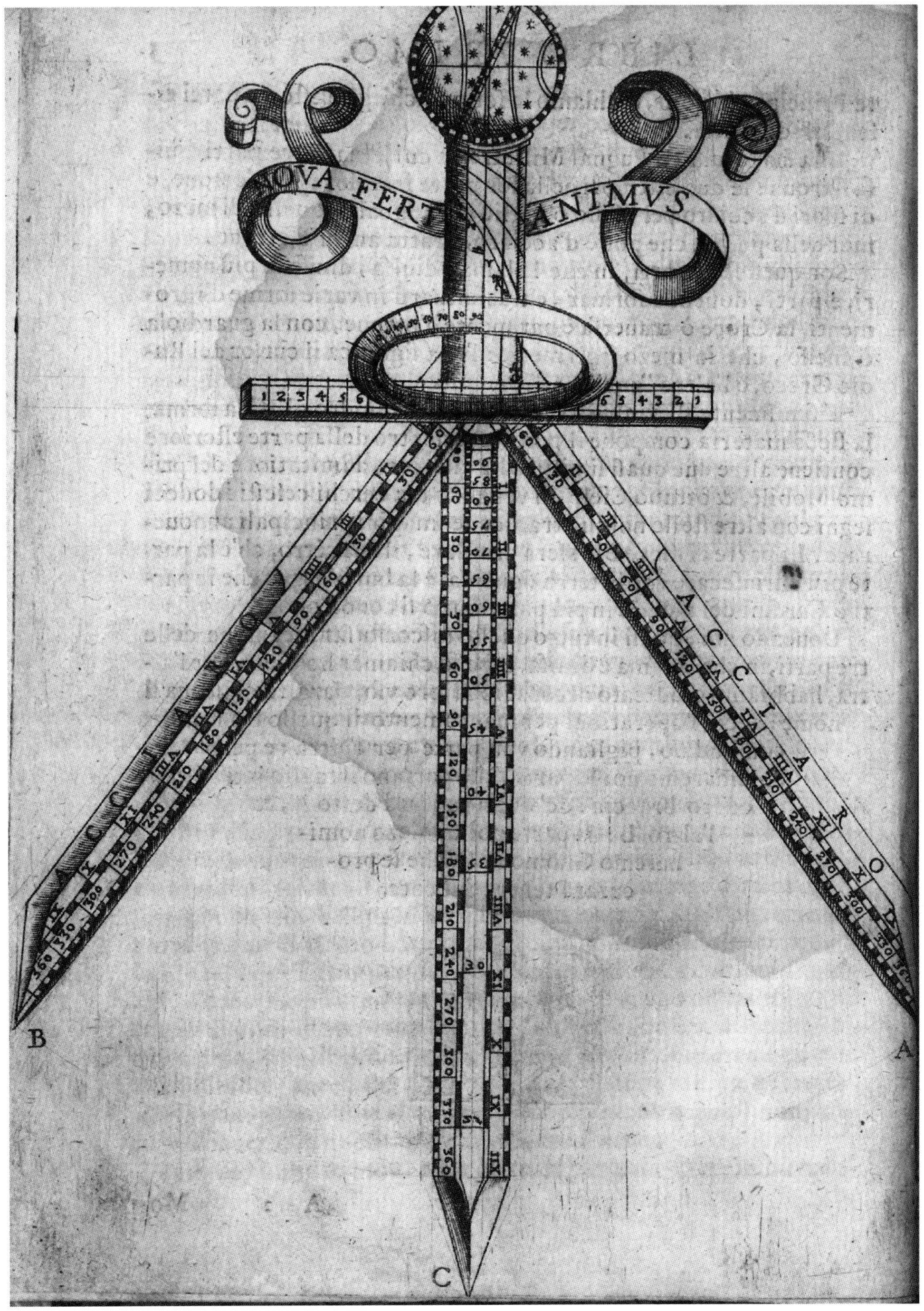

12. Bartolomeo Romano, *Proteo militare* (Naples, 1595).

Although many mathematical instruments of the Renaissance sought to combine a range of observational and calculating functions in one device, few (if any) achieved the selfconscious universality of the 'military proteus'. The name was deliberately chosen to highlight the instrument's capacity for transformation into a wide variety of configurations.

The first of the volume's three books deals with the construction of the instrument and its transformation into other instruments such as the 'radio latino' (catalogue nos 59 and 60). The second and third books embrace the various contexts in which the 'proteo' can be used, ranging across perspective, architecture, cartography, sculpture, fortification, typography, surveying, mensuration, astronomy, dialling and navigation.

As befits an instrument which was designed in the form of a dagger, there are also military uses of the 'proteo'. It is shown acting as a gunner's quadrant, and was intended to provide assistance both in ordering troops in formation and in setting out a military encampment. With the appropriate scales, it could also be used as a gunner's gauge. Further material on gunnery appears in a discussion of the design of various artillery pieces. The author, who published his *Nautica mediterranea* (1602) on navigation and shipbuilding under the name Bartolomeo Crescentio, displays his naval interests by mentioning the stone-throwing perrier guns used on board Venetian galleys.

This copy of the book is missing its title page and dedication.

13. Leonhard Zubler, *Nova geometrica pyrobolia. Neuwe geometrische Büchsenmeisterery* (first edition, 1608; this edition, Zurich?, 1614).

Leonhard Zubler's 'geometrical gunnery instrument' did not have the pretensions to universality of the 'proteo militare' but was still expected to fulfil a range of roles. Like several other contemporary devices (catalogue nos 7, 60 and 64), Zubler's instrument had two jointed legs attached to a central leg, with smaller supporting limbs providing additional links between the central and side legs. In Zubler's version, the supporting limbs slide in a slot in the central leg when the angle between the side legs is altered. The slider moves against a scale of degrees to indicate the angle so that, with a plumb bob and line, the instrument can take the place of a gunner's quadrant. Gunner's gauge and inch scales enable the device to act as a pair of measuring calipers, and Zubler discusses the size and weight of shot and quantity of powder for both long guns and mortars, as well as the ranges they achieved.

As with Zubler's other tracts on new instruments (e.g. catalogue nos 54 and 58), the volume is liberally illustrated with finely executed copper-plate engravings. Zubler was well aware of the visual appeal and distinction of his publications and typically advertised the engravings on his title pages – in this case we are told that the book comes with 'schönen kunstreychen Kupfferstucken'. Among the illustrations are several showing gunner's sights in use. The sights are mounted at the breech end of cannon and have a sliding pinhole moving in a slot against a scale. The gunner set the required elevation on the sight and then adjusted the inclination of the gun until the pinhole, target and top of the gun's muzzle came into alignment.

Although the focus of the text is on gunnery, the book's final section deals with the more general uses of the instrument, especially in mensuration and surveying. To aid in such work, the instrument carries a compass. Zubler also mentions in passing that the instrument can in principle be used for astronomical observations.

At the end of the book is an advertisement for the instruments available from Zubler, who lived and worked in Zurich. Artillery instruments are grouped along with sundials, astrolabes, quadrants, nocturnals and sectors, indicating that gunnery instruments were here firmly identified as mathematical instruments.

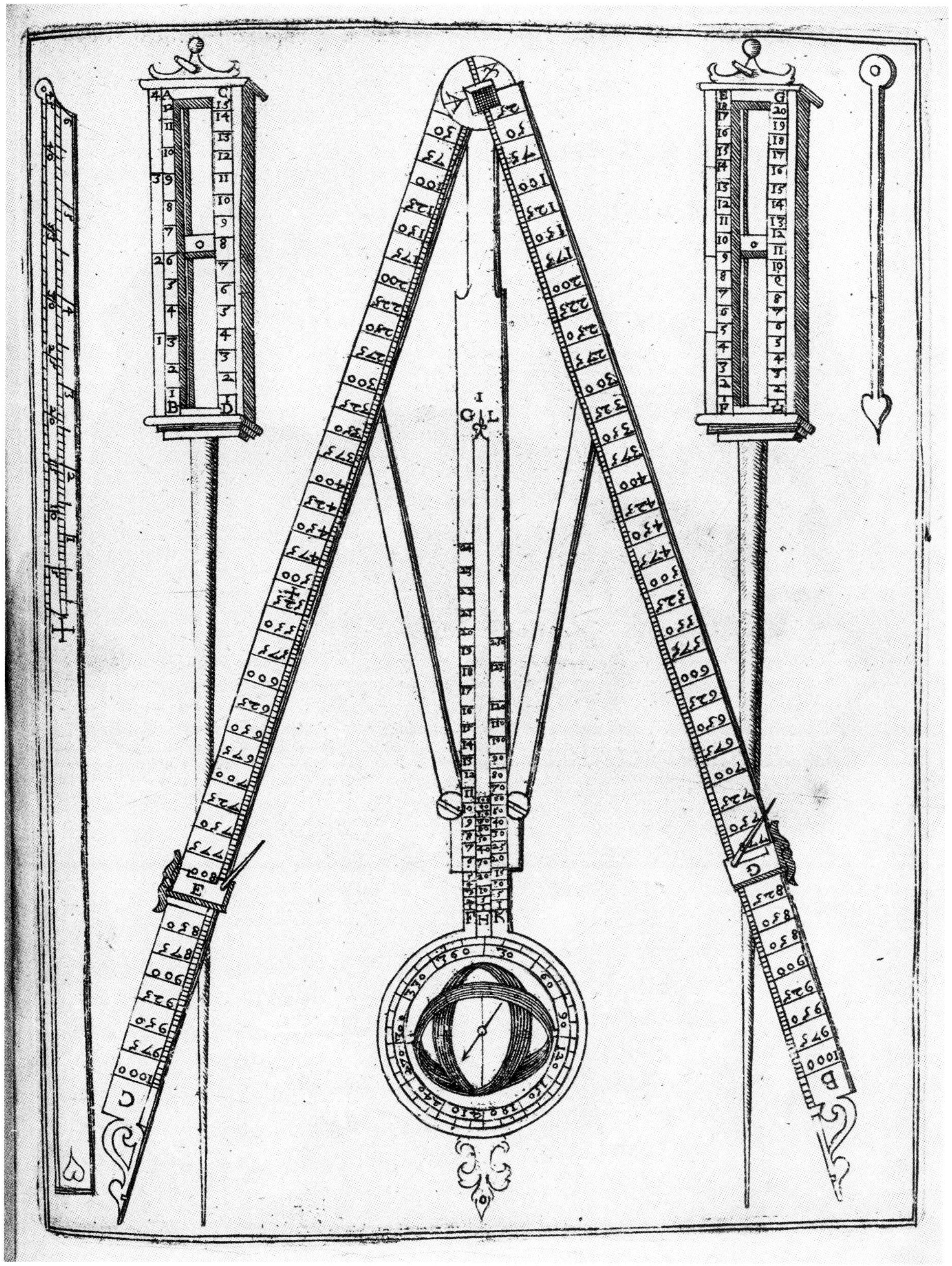

NOVA GEOMETRICA PYROBOLIA

Neuwe
Geometrische Büchsenmeisterery:
Das ist:
Grundtlicher Bericht / wie man
durch ein neuw Geometrisch Instrum:nt/
mit sonderer behendigkeit/jedes Geschütz klein oder
groß/bey tag oder nacht nit allein richten/sonder
zugleich auch desselben höche vnnd weite
mässen soll:
Mit schönen kunstreychen Kupfferstucken
geziert/vnd anjetzo erstlich/ allen Kunstliebenden/
sonderlich aber Zeüg vnnd Büchsenmeiste-
ren zu gutem an Tag gegeben/
durch
Leonhard Zubler/Burger
zu Zürich.
Mit Röm. Keys. May. freyheit/ nicht
nach zutrucken.

Basel. M. DC. XIIII.

Fig. 27 (left)
Engraved plate showing Leonhard Zubler's 'geometrical gunnery instrument'.
From catalogue no. 13.

Fig. 28 (right)
Title page from Zubler's book on his 'geometrical gunnery instrument'.
From catalogue no. 13.

14. Surveying and Gunnery Instrument
c. 1600

This instrument is very close to the design of Zubler's 'geometrical gunnery instrument'. It differs principally in having shorter side legs, so that it cannot be used as a pair of large dividers or calipers. The joint linking the side and central legs is also different: whereas Zubler shows all three limbs mounted together on the same axis, in this instrument the two side legs are held in place by blued-steel spring blades.

The central leg carries a double scale of polygons and a scale of degrees. As in Zubler's depiction there is a compass at its end, but in this case it is provided with a string-gnomon dial for about 48° latitude. There are also scales for Prague, Rome and Vienna inches.

Another instrument of this type, now in Florence, is signed 'I. B. F.', possibly an abbreviation for 'Iost Bürgi fecit' (Miniati pp. 32–33). In the quality of their construction, materials and engraving these two instruments are certainly compatible with other instruments by Bürgi (cf. catalogue no. 52).

Length: 356 mm
Inventory no. 47,493

15. Gunnery and Surveying Instrument
16th century

The square format of this instrument is unusual and includes several different components, several now incomplete. Among its operations, artillery uses are most strongly signalled by the small vignette of a firing gun incorporated into its engraved decoration.

The instrument's central slit has a scale of inches and would once have housed a sliding gunner's sight. Another missing piece is the plummet which hung to the side of the sight. There is a table for converting between feet (SHV – Schuh) and paces (SHRI – Schritt), such as is found in range tables like those published by Leonhard Zubler (catalogue no. 13). The table correlates with another on the other side of the instrument. Of further use to gunnery, though for more general surveying purposes too, are a shadow square and quadrant, along with two sighting rules. A pivoted compass and a sundial (lacking its gnomon) are both designed to be used with the square plate upright; the necessary stand is now missing.

Dimensions: 143 x 143 mm
Lewis Evans Collection
Inventory no. 46,878

16. Gunner's Sight
17th century

An unusually long gunner's sight – graduated to 9½ inches – which enables a gun to be set to a high elevation. To improve the instrument's portability it separates into two parts (just above the 5½ inch line); the upper part is held in place by a spring catch. Rather than a slider in a central slot, the sight is attached at the side of a sliding cursor. The short projection on the underside of the stand enables alignment with the centre line of the gun's barrel.

Length: 266 mm
Inventory no. 41,377

Fig. 29 (above)
An artillery scene as instrumental detail.
Catalogue no. 15.

Fig. 30 (right)
Gunner's sight.
Catalogue no. 16.

Catalogue no. 15 is also illustrated in colour on page 37 and catalogue no. 14 on page 40.

Catalogue no. 17 is illustrated in colour on page 41.

17. Gunnery Instrument

Signed: Pragae fecit Erasmus Habermel
Late 16th century

Erasmus Habermel excelled in the construction of elaborate instruments whose ingenuity of form was matched by their fine gilding and decoration. He typically produced devices for a courtly clientele who valued the spectacular display of practicality over the utility of working instruments. However, unlike his astronomical compendium with string-gnomon dial (catalogue no. 5), which incorporates gunner's gauge scales almost as a decorative afterthought, this instrument is all but exclusively devoted to gunnery. It is one of several hybrid artillery instruments by Habermel to survive.

The instrument combines three distinct devices: a gunner's quadrant, a sight and a gauging rod. The quadrant, when seen stripped of other components such as the stand and sight, is of disarming simplicity and closely matches the form illustrated by Tartaglia. Its plumb bob and line are set against an arc graduated in points rather than degrees, and its long leg would have been inserted into a gun's muzzle in use. Two small sights attached to the side of this long leg enable the quadrant to be used for more general observations. The leg also carries the standard gauge scales for determining the weight of iron, lead and stone shot.

To transform the instrument into a sight that can be placed on the breech of a gun, the quadrant is turned upside down and made to serve as no more than a frame, to which the stand and sight attachments are screwed. The slot for the sliding sight is graduated not only in numbers from 0 to 12, with subdivisions to eighths, but also using letters from A to M marked against the same units.

The quadrant and sight share no common structural features but their combination creates an impressive and elaborate spectacle. Habermel's instrument is a *tour-de-force* whose primary purpose was presumably to grace the collection or cabinet of a noble or wealthy patron.

Dimensions: 350 x 194 mm
Billmeir Collection
Inventory no. 41,591

18. Surveyor's Quadrant
Signed: Iacobus Lusuerg Mutinensis Faciebat Romae
A° 1677

If Erasmus Habermel's combined quadrant, sight and gauge was devoted to the ostentatious demonstration of aids to gunnery, this quadrant by Lusverg is a model of discreet restraint in comparison. Nevertheless its prospects for active military service were probably no higher.

The upper face of Lusverg's instrument is marked out for a surveyor, with folding sights and a sinical quadrant. However, in the otherwise blank space alongside the ball and socket joint on the under side, there is a circular scale of points marked 'Pro Eleuatione Bombardæ'. To measure the elevation of an artillery piece, a plumb bob and line are attached through a hole pierced directly through the plate of the instrument. To forestall any uncertainty, an engraved diagram gives visual instruction on how to use the instrument at the mouth of a gun.

Radius: 157 mm
Inventory no. 48,133

Fig. 31 (top)
German gunner's level and gauge.
Catalogue no. 19.

Fig. 32 (above)
Surveyor's quadrant with gunner's level.
Catalogue no. 18.

19. Gunner's Level and Gauge
Early 18th century

The long leg of this instrument enables it to be placed in the barrel of a gun to read elevations against a scale of degrees from 45–0–45. The rigid plummet is intended to give quicker readings than were possible with a plumb bob and line, which were liable to continue swinging for longer. The instrument also serves as a gunner's gauge, carrying scales for stone, iron and lead on one side, and scales for Vienna and Nuremberg inches and for powder on the reverse.

Length: 296 mm
Inventory no. 44,944

Catalogue no. 20 is illustrated in colour on page 37.

20. Gunner's Level and Sight, with Sundial and Compass

Signed: VDALRICVS KLIEPER AVGVST FACIEBAT ANNO LXXVIII
Dated: 1578

Although the earliest quadrants and levels were designed to be inserted into the muzzle of a gun, the gunner who did so risked exposure to enemy fire. Instruments which indicated the elevation of a gun at the breech end were less dangerous for their users. Many such levels survive, in a great range of forms and styles. Even when they were undoubtedly not intended for active service, breech levels had another advantage: they could readily be incorporated into combination instruments. This elaborate and finely gilt instrument is just such a combination: not only does it juxtapose several gunnery functions but its hinged upper section carries a horizontal sundial and a small compass in addition to a decorative plummet.

The main upright plate of the instrument has its own plummet with a short levelling arc, supported on a hinged leaf which can be swung round to either side of the upright. When folded out, the level reveals a table lettered in red which provides data on shot and powder for various types of artillery. The upright also carries an accompanying graphical table and has a central slit for a sight which is now missing. Like Erasmus Habermel's gunnery instrument (catalogue no. 17) the sight scale is graduated not only with numbers (in this case for inches) but also letters from A to M. The whole upright sits in a graduated slot in the arched foot and can be laterally adjusted, with its position fixed by two screws. The foot itself has two hinged end-pieces which raise or lower the instrument.

Taken with its tooled, leather-covered case, the instrument has clearly been crafted for visual appeal. Ulrich Klieber, its maker, is primarily known as a clockmaker working in Augsburg, but several gunnery instruments signed by him survive.

Height: 156 mm
Inventory no. 35,522

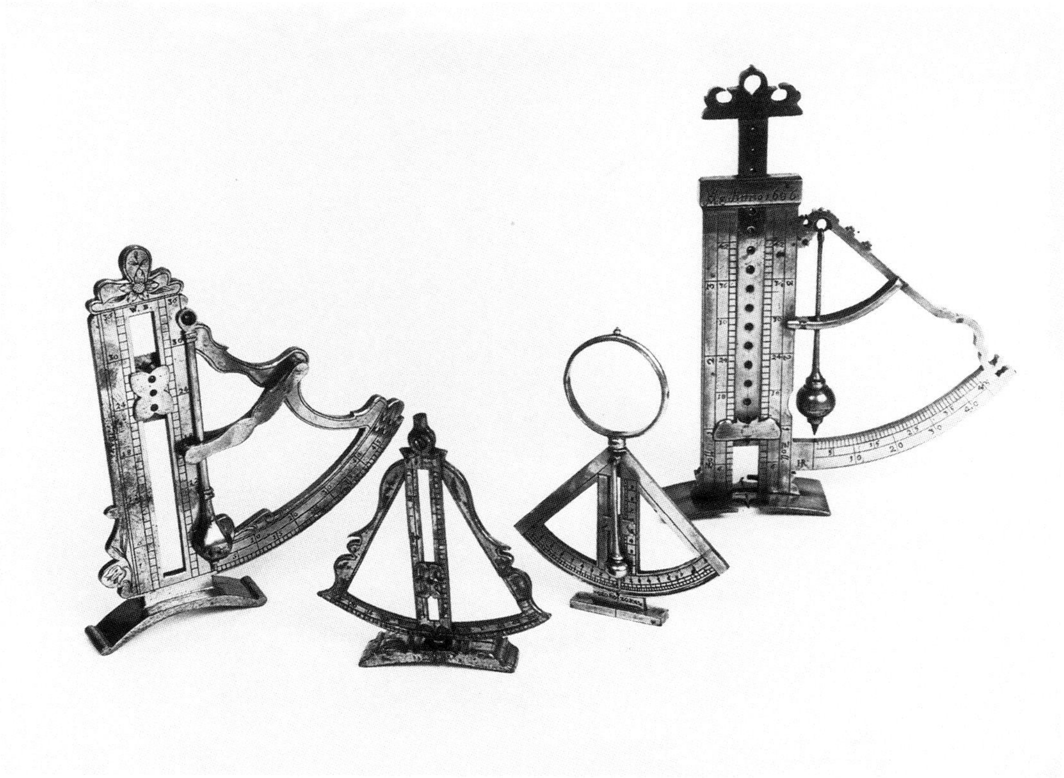

Fig. 33 (above)
A group of gunner's levels and sights.
From left to right, catalogue nos 30, 22, 21 and 24.

21. Gunner's Level and Sight
Signed: .GEORG.ZORN. .1.6..2 4.

The most common surviving gunnery instruments are not complex devices such as Klieber's ornate confection but combined sights and levels. Georg Zorn's quadrant level is a typical example, with a rigid plummet and sliding sight. In use as a level, the instrument would be set up near the breech of the gun. With the brass quadrant aligned with the gun's axis the plummet reading on the levelling arc would then be read from the side. On the reverse of the instrument the scale of the levelling arc is graduated in points of the gunners quadrant, with equal parts in the range 6–0–6. The arc scale on the front is graduated in unequal parts 9–0–9 and is meant to be used with a particular type of artillery. Likewise, the sight operates not with universal units but two particular scales, from 1–10 on the front and 1–7 on the reverse. When used as a sight, the instrument would be placed transversely across the gun barrel. Rather than an arched or angled foot, to ease transverse placement on a gun barrel, the instrument has a flat base whose underside carries a subdivided inch scale.

The maker Georg Zorn was based in Augsburg and amongst his production were more elaborate artillery instruments such as Zubler's 'geometrical gunnery instrument' (catalogue no. 13; Dreier, p. 111; Zinner, p. 597).

Height: 88 mm
Inventory no. 38,318

22. Gunner's Level and Sight
c. 1600

Instruments such as Zorn's, which combine a level and sight, often have the inconvenience that the plummet obscures the sighting holes when the instrument is level. In this anonymous example, the problem is overcome by suspending the sight holder freely and dispensing with a separate plummet. The pointer at the end of the sight holder moves over an arc of slightly more than 60° whose scale of equal parts runs 24–0–24. The divisions of the sight scale correspond to Nuremberg inches, and are subdivided to eighths with the half-inch points marked. Unlike many levels, the arc simply screws into the arched foot rather than being formed from a solid piece of brass. The instrument would once have been surmounted by a ring, as in the level and sight by Georg Zorn (catalogue no. 21).

Height: 75 mm
Inventory no. 46,988

Fig. 34 (top left)
Instruments with gunner's gauge scales.
Catalogue nos 4, 5 & 6.

Fig. 35 (top right)
Surveying and gunnery instrument supported by Tartaglia.
Catalogue nos 1 & 15.

Figs 36 & 37 (right)
Gunner's level and sight, with sundial and compass.
Catalogue no. 20.

23. Gunner's Level and Gauge
Signed: F. F. F. 1629.

This instrument has the smallest level and plummet of all those in the exhibition: the radius of the arc is only 35 mm. The plummet's support is nevertheless finely formed and engraved in silvered brass. The arc scale covers about 73° and is divided in equal parts 15–0–15.

The accompanying stiletto gauge is more appropriate to a goldsmith than a gunner. It carries half ounce ('Lot') scales for spheres of gold, silver, lead and tin rather than the familiar iron, lead and stone. In addition, it is calibrated only up to one pound for each of the metals, whereas gunner's gauges typically indicate the weight of much heavier shot. The gauge is attached to the back of the level by two screws. These are individually identified with their sockets by single and double dots respectively. By providing such 'instructions' for securing the gauge, the maker evidently envisioned a user detaching and reattaching it at will, reinforcing the implication that the instrument's two components were used separately.

The maker should probably be identified as Franz Viebig, who signed an elaborately decorated gunner's level as 'Franciscus Fiebig me faciebat Arnstadiae Anno 1633' (Dreier, p. 108; Zinner, p. 570).

Height: 161 mm
Inventory no. 37,342

Fig. 38 (top)
Gunner's level and gauge, 1629.
Catalogue no. 23.

Fig. 39 (bottom)
Attachment of gauge with socket-matched screws.
Catalogue no. 23.

Catalogue no. 24 is illustrated on page 36; catalogue nos 25 and 26 on page 44.

24. Gunner's Level and Sight

Signed: [crown] M:E: Anno 1686. [also signed] ĦR

Rather than having a single sighting hole sliding in a slot, this instrument has a long slider pierced with a sequence of seventeen holes. Only slight adjustments of the slider are therefore necessary to set a hole against a particular elevation. Settings are read off from a scale of inches subdivided to twelfths. The level is offset from the sight, to avoid obscuring it, and has a brass plummet moving over a 45° arc. By the later 17th century it was becoming more common to graduate levelling arcs using the degrees of the 'astronomical quadrant' than to use scales appropriate to only one particular gun or variety of shot.

Height (slider not extended): 125 mm
Inventory no. 42,501

25. Gunner's Level and Sight

Signed: MK MDCLXXXIX
Dated: 1689

When used as a level, the rigid plummet of this gilt brass instrument moves over a quadrant scale graduated in degrees 45–0–45. Sighting is accomplished through a series of nine small holes pierced through the quadrant plate. The holes are at ¼ inch intervals and sighting observations could therefore only have been approximate. By contrast, the scale on the quadrant arc is continuous and so should permit finer readings. But its single degree divisions are marked irregularly and inaccurately.

The shaped, leather-covered case for this small gilt brass instrument also survives.

Height: 72 mm
Inventory no. 49,981

26. Gunner's Level

Signed: Butterfield ÆParis
c. 1700

Not all pocket-sized instruments incorporated both a level and sight. The observing part of this instrument by the Paris-based English maker Michael Butterfield consists only of a level. The decorated steel plummet moves over a scale of 45° whose arc is hinged to fold into alignment with the upright. As Nicolas Bion noted when describing and illustrating this form of instrument, the folding construction simply made the instrument more compact and easier to fit into a case (catalogue no. 32).

The reverse of the upright carries gauge scales. There is a joint scale for 'Calibres des Pieces' and 'Diamettres et Poids des Boulets' which allows for windage and provides the weight of shot from ¼ to 5 pounds. There is also a scale for powder calibrated over the same range and marked 'Demy Diamettres des Spheres des Poudres'.

Height: 102 mm
Inventory no. 52,788

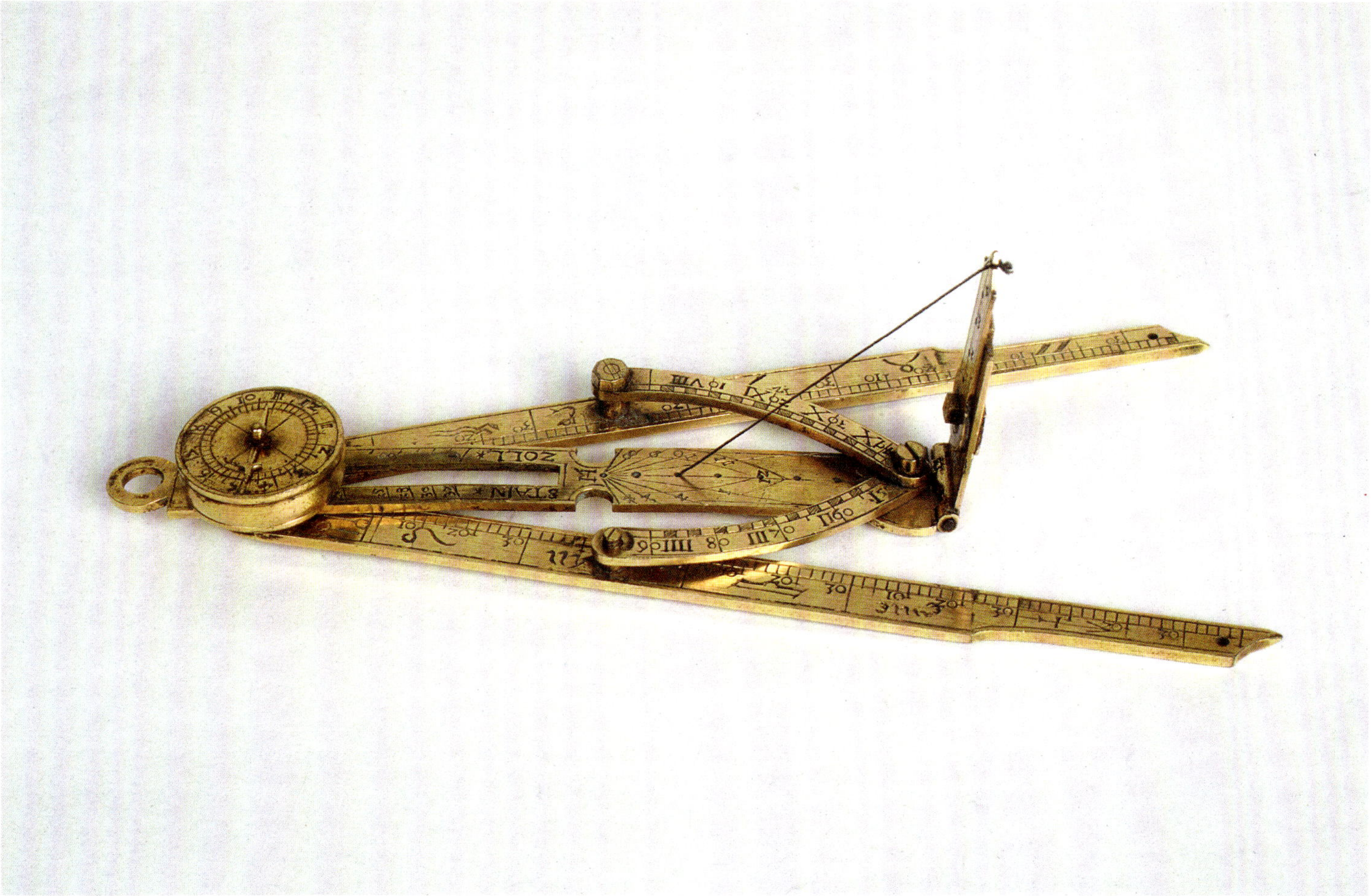

Fig. 40
German gunnery and dialling instrument, 1595.
Catalogue no. 7.

Fig. 41
Surveying and gunnery instrument, *c.* 1600.
Catalogue no. 14.

Fig. 42 (below)
Gunner's compendium by Erasmus Habermel.
Catalogue no. 17.

Fig. 43 (right)
Gunner's quadrant from the Habermel compendium.
Catalogue no. 17.

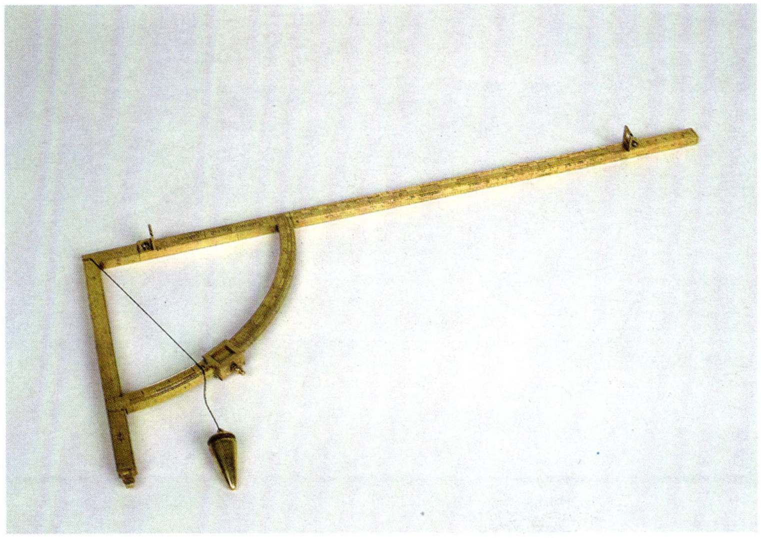

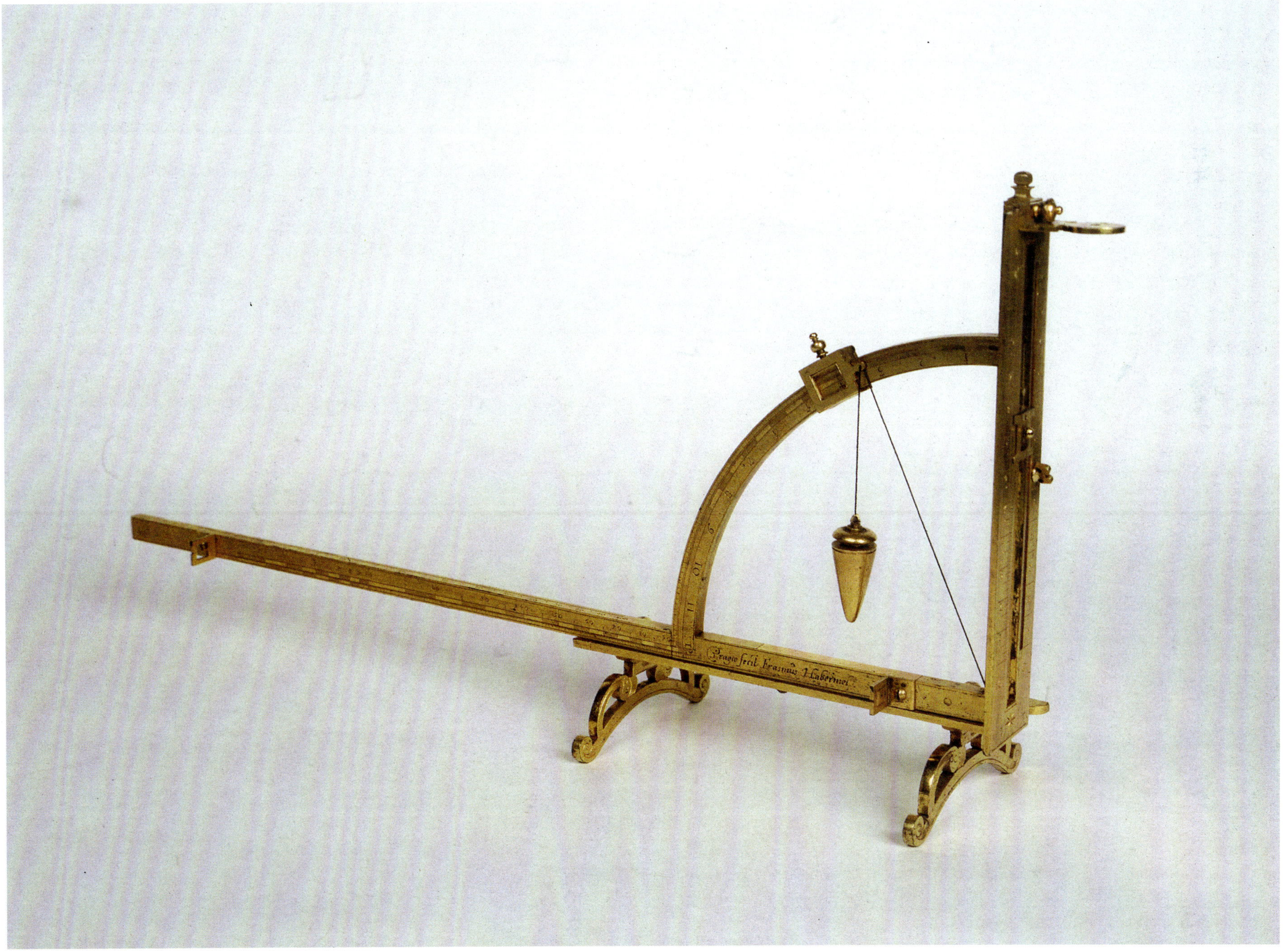

27. Gunner's Level and Sight

Signed: Picart A Cambray
c. 1700

Unlike German examples of combined levels and sights (such as catalogue nos 24 and 30), the offset levelling arc of this French instrument uses a solid sector of metal. However, it is otherwise similar in form, with a single sight sliding against a scale graduated 5–45 and the plummet moving over a 45° scale. There are two additional sights, one fixed at the foot of the sighting upright, and the other at the end of the levelling arc. In addition to the maker's inscription, the reverse of the instrument also has the engraved name of 'J. WEILER', perhaps an owner of the late 19th or early 20th century.

Height: 110 mm
Inventory no. 34,210

28. Gunner's Level

c. 1700

The turned plummet is suspended by a hook and ring and moves over a 45° scale. Both the upright and the base carry engraved decoration. The brass bar which is screwed to the underside of the level may be the cut-down remainder of a square-sectioned foot rule with gunner's gauge scales: one of the visible sides is graduated in inches from 9–12, while the other two exposed sides carry partial scales with the ranges 70–125 and 20–50. The instrument has a leather-covered case.

Height: 103 mm
Inventory no. 34,455

29. Gunner's Level

Early 18th century

This level has a quadrant arc divided in degrees 45–0–45 and a rigid plummet with an engraved bob.

Height: 124 mm
Inventory no. 54,544

Fig. 44 (above)
Gunner's level.
Catalogue no. 29.

Catalogue no. 27 is illustrated on the next page; catalogue no. 28 in colour on page 44; and catalogue no. 30 on page 36.

30. Gunner's Level and Sight

Signed: W.B. and W: BURUCKER. NORI: [Nuremberg]
Early 18th century

As with other combined gunnery instruments in the exhibition this German instrument has a levelling arc offset from the upright of a sight. The sight moves in a slot against a scale of equal parts from 0–36 while the plummet moves over a scale of degrees from 0–45 and is constrained by a guard. The instrument still has its original leather-covered wooden case.

The maker Burucker was active in Nuremberg in the first half of the 18th century and is also known to have produced sundials (Zinner, p. 277).

Height: 117 mm
Inventory no. 47,357

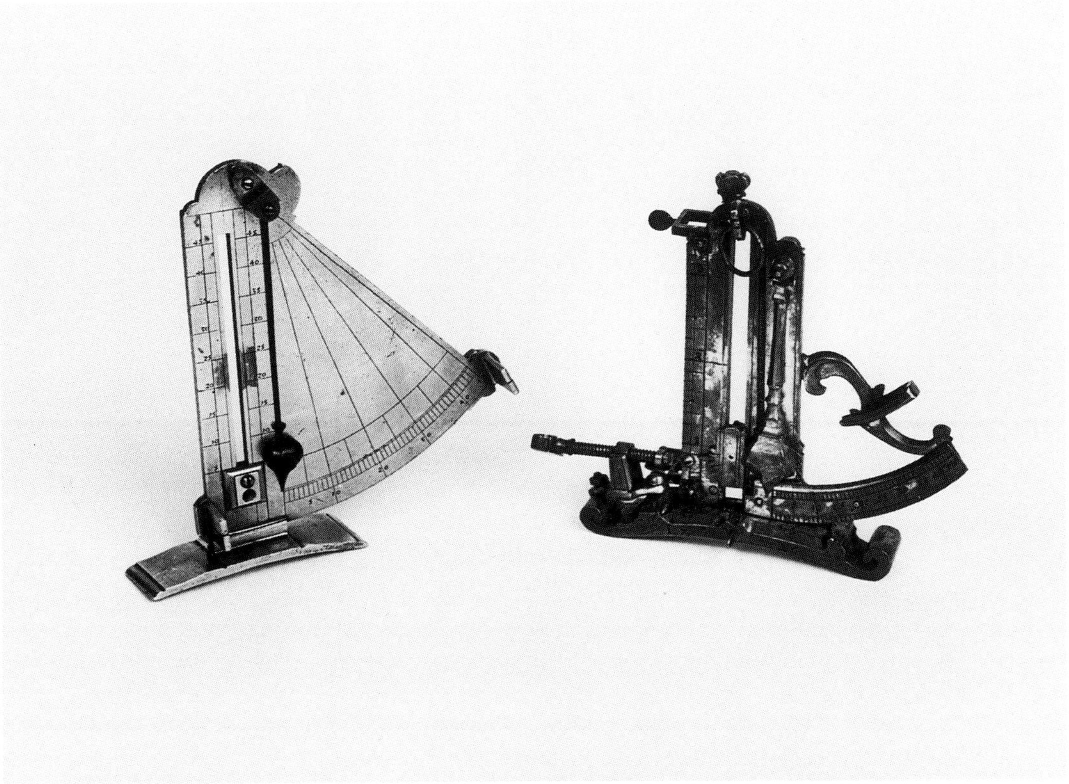

Fig. 45 (above)
Two gunner's levels and sights.
Catalogue nos 27 & 31.

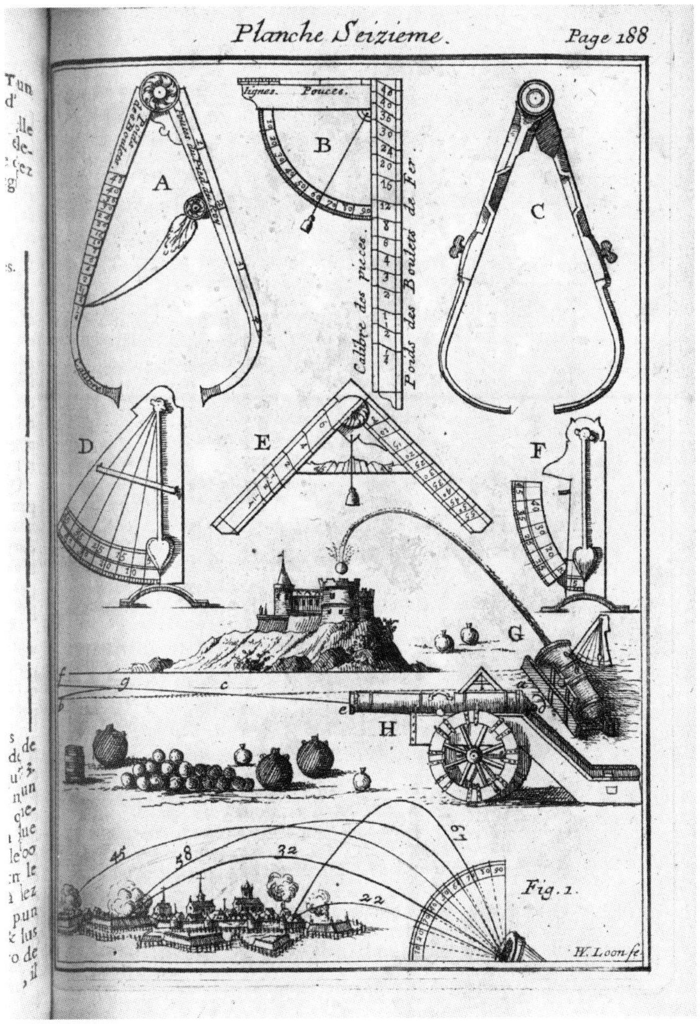

Fig. 46 (right)
Gunnery instruments illustrated by Nicolas Bion.
Catalogue no. 32.

31. Gunner's Level and Sight
Dated: 1736

By the second quarter of the 18th century the format of an offset 45° arc and vertical sliding sight was evidently well-established for the gunner's combined level and sight. This instrument adds some refinements by using two screw threads, one to adjust the position of the sight and the other to move the upright across the foot. In addition, the reverse of the instrument carries two square brackets which could be used to hold a vertical device, perhaps a perpendicular to mark the sight line on a barrel.

Height: 105 mm
Inventory no. 51,507

32. Nicolas Bion, *Traité de la construction et des principaux usages des instrumens de mathematique* (Paris, 1709).

Bion's text was intended to provide a survey of the full range of mathematical instruments then available, and it included a section on gunnery. His well-illustrated work served as a standard reference not just through its several 18th-century French editions, but on a more international stage too. A German edition was published in 1712 and frequently reissued and an English translation by Edmund Stone appeared in 1723. Stone's edition is particularly interesting for it adds specifically English instrument types absent in the French original. Thus a description and engraving of English gunner's calipers (see catalogue no. 10) were added to the account of French gunnery instruments. The second English edition of 1758 further supplemented the original offering, in effect proclaiming the superiority of the latest English innovations.

In the original text Bion described the construction and use of various calipers and levelling instruments for guns. In addition to levels using rigid plummets moving over an arc scale (such as catalogue no. 26), he shows a folding square set up on a breech (catalogue nos 33 and 34). His engraved plate also gives readers unacquainted with warfare a general sense of the different trajectories typical for cannons and mortars and provides a graphical display of the variation of range with elevation.

Although Bion is now probably best known as an author he was equally recognized as an instrument maker in his own day (cf. catalogue no. 9). He achieved an official position as royal instrument maker, with the title 'ingénieur du roi pour les instruments de mathématique'.

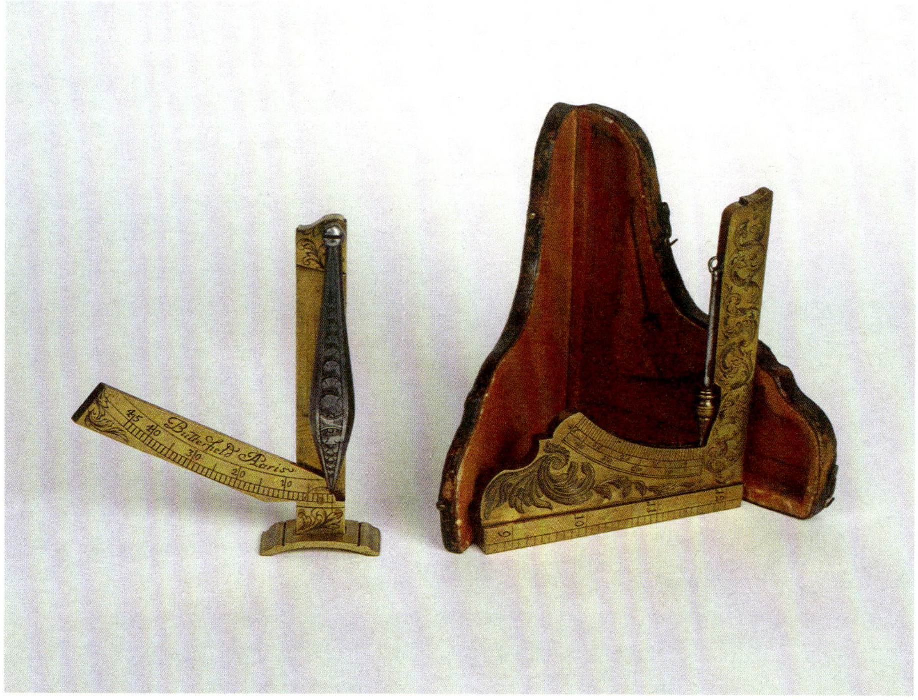

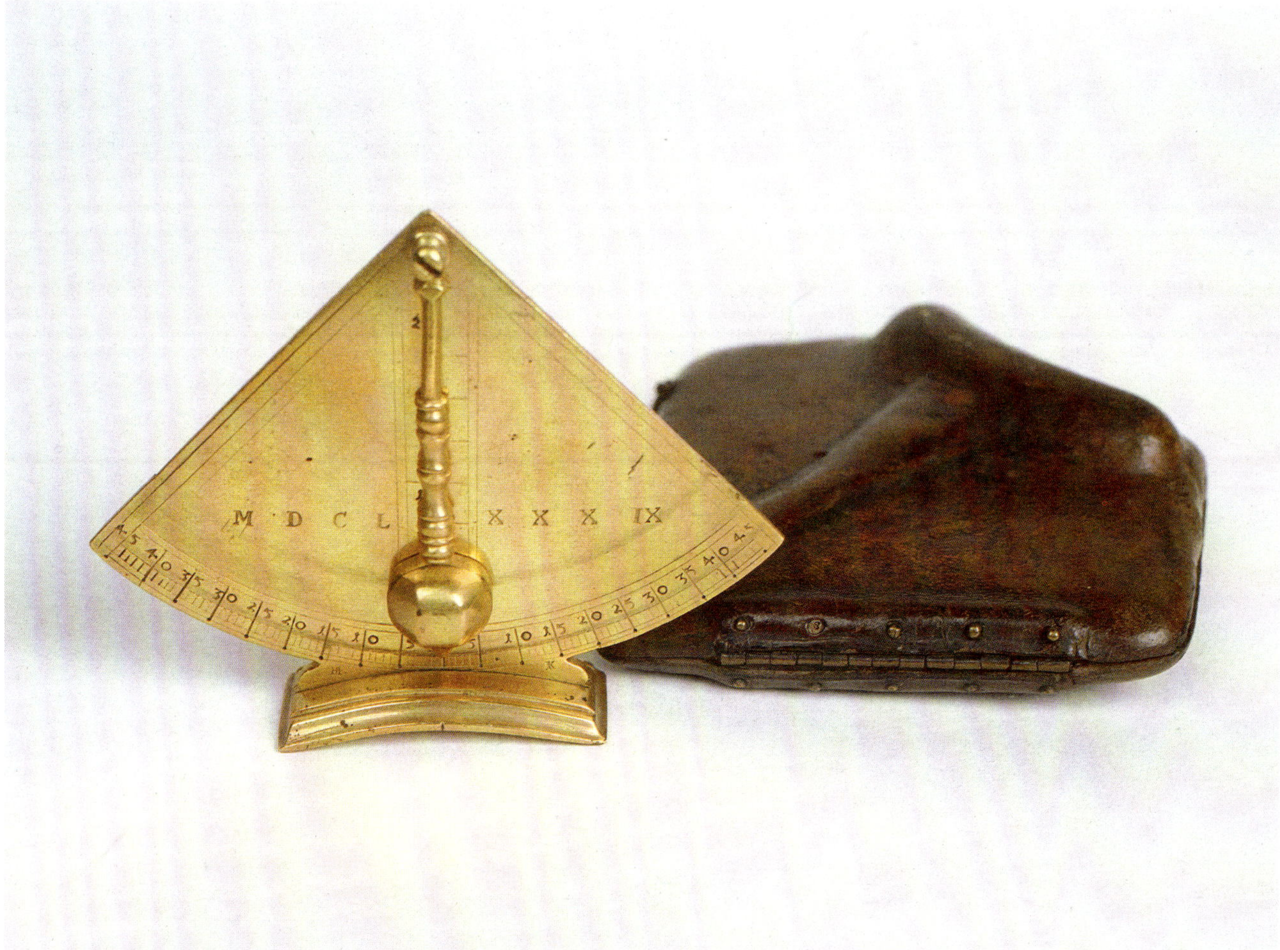

Fig. 47 (top left)
Two gunner's levels, one with a folding arc, the other fixed.
Catalogue nos 26 & 28.

Fig. 48 (above)
English gunner's perpendicular by Adams, *c*. 1790.
Catalogue no. 36.

Fig. 49 (left)
Gunner's level and sight, 1689.
Catalogue no. 25.

Fig. 50 (above)
Two French levelling sectors and an English folding square.
Top to bottom, catalogue nos 33, 34 & 35.

33. Folding Level and Sector
Signed: Le Maire Lefils ÆParis
c. 1750

An instrument of this type is shown by Bion mounted for levelling a gun. A plumb line is attached at the hinge and reads against a scale on the cross-piece. Although identical in form to the Bion instrument, this example by Le Maire *fils* has only a single line on the cross-piece to indicate when a horizontal level has been achieved. It also diverges from Bion in the scales engraved on the legs. Bion describes an instrument solely for gunnery which carries scales for the diameter of shot and the calibre of guns. Le Maire has included not only these scales but also more general-purpose sectoral scales of chords, solids, metals, equal parts, planes, and polygons.

Length (closed): 182 mm
Inventory no. 42,231

34. Folding Level and Sector
Signed: Clerget A paris
c. 1700

Clerget's instrument is simply a miniature version of the Le Maire rule and may originally have been supplied as part of a pocket set of silver drawing instruments. Two of the sectoral scales (planes and solids) have been squeezed out due to its smaller size but the other features remain the same, and there is even an extra engraved scale of inches on the outside edge.

As with the larger instrument, the cross-piece is used to fix the instrument as a square for levelling. The cross-piece is attached to one of the two legs and can slide along a groove in the other. A catch in the groove holds the cross-piece when the legs are at right angles to each other. The two legs are bevelled at the ends to give greater stability when it is set up as a square.

Length (closed): 88 mm
Inventory no. 43,341

35. Folding Square
Signed: I: Marke fe.
c. 1670

The folding rules by Le Maire and Clerget represent a standard French pattern of instrument, of which many examples survive. This English folding square is an earlier and altogether more unusual variant on the theme. The instrument has two principal legs with a maximum opening of 90°. When the right angle has been achieved the hinged supporting struts lock into a straight line. For levelling, a plumb line can be attached through a hole at the centre of the main hinge and read against a scale of degrees on the support, which also carries sectoral scales, including one for polygons. The principal legs have a range of equal parts and trigonometrical scales.

Length (closed): 182 mm
Inventory no. 39,708

36. Gunner's Perpendicular

Signed: Adams London

c. 1790

Catalogue no. 36 is illustrated in colour on page 44.

A perpendicular enabled a gunner to establish the centre line of an artillery piece and thus its direction of fire. The instrument was placed transversely on the gun's barrel and, once levelled using the spirit level, the steel plunger was depressed. Marks were made at the muzzle and breech of the gun and joined by a chalk line. Perpendiculars incorporating spirit levels were first purchased by the English Office of Ordnance in 1757 and replaced earlier versions in which the levelling was carried out with a plumb bob and line (Millburn 1995, p. 35).

The instrument has a shaped, fishskin-covered case of wood and pasteboard.

Height: 115 mm
Inventory no. 51,485

37. Daniel Santbech, *Problematum astronomicorum et geometricorum sectiones septem* (Basel, 1561).

All the many varieties of sights and levels only had a practical purpose if the gunner knew the relation between elevation and range for the guns under his command. From Tartaglia's first publication on artillery onwards, such tables of 'randons' were promised and printed. Yet the precise nature of the shot's trajectory remained a thorny issue for mathematicians from the 16th to the 18th centuries, and beyond. Tartaglia had adapted the terms of academic natural philosophy to offer his 'new science' of artillery and Galileo announced the parabolic nature of the projectile's motion in his own 'new science' of falling bodies. Attempts to incorporate the problematic effects of air resistance led to increasingly sophisticated mathematical treatments in the later 17th and 18th centuries, but these rapidly outstripped the reach of ordinary gunners.

Accounts of ballistics based on Tartaglia remained standard in gunnery manuals long after alternative notions were available. However his views were not uniformly adopted by early practitioners. Some authors, such as father and son Leonard and Thomas Digges (catalogue no. 45), explicitly rejected Tartaglia's ideas. Other authors, such as Daniel Santbech, seem not to have been aware of his writings.

Santbech's discussion shows how quickly artillery became a subject of learned discourse as well as practical art, since his account appears in a more general mathematical text encompassing astronomy, sundials, surveying, and levelling for water courses. He sets out to investigate gunnery theoretically on the firm foundations of geometry, quoting Euclid and Ptolemy in the Greek, while using the geometrical quadrant discussed elsewhere in his text as the necessary instrumental accompaniment.

At first glance his illustrations of straight-lined trajectories with abruptly acute angles appear absurd. However, they are depicted for calculating convenience rather than graphic realism. Santbech acknowledged that a cannon ball would not fly in a straight line and then suddenly fall directly to the ground. He substituted a straight line for the actual path of the shot so that he could create a right-angled triangle from which ranges were computed with the help of a table of sines.

Figs 51 & 52
Daniel Santbech's
mathematical treatise:
title page and illustration
of artillery trajectory.
From catalogue no. 37.

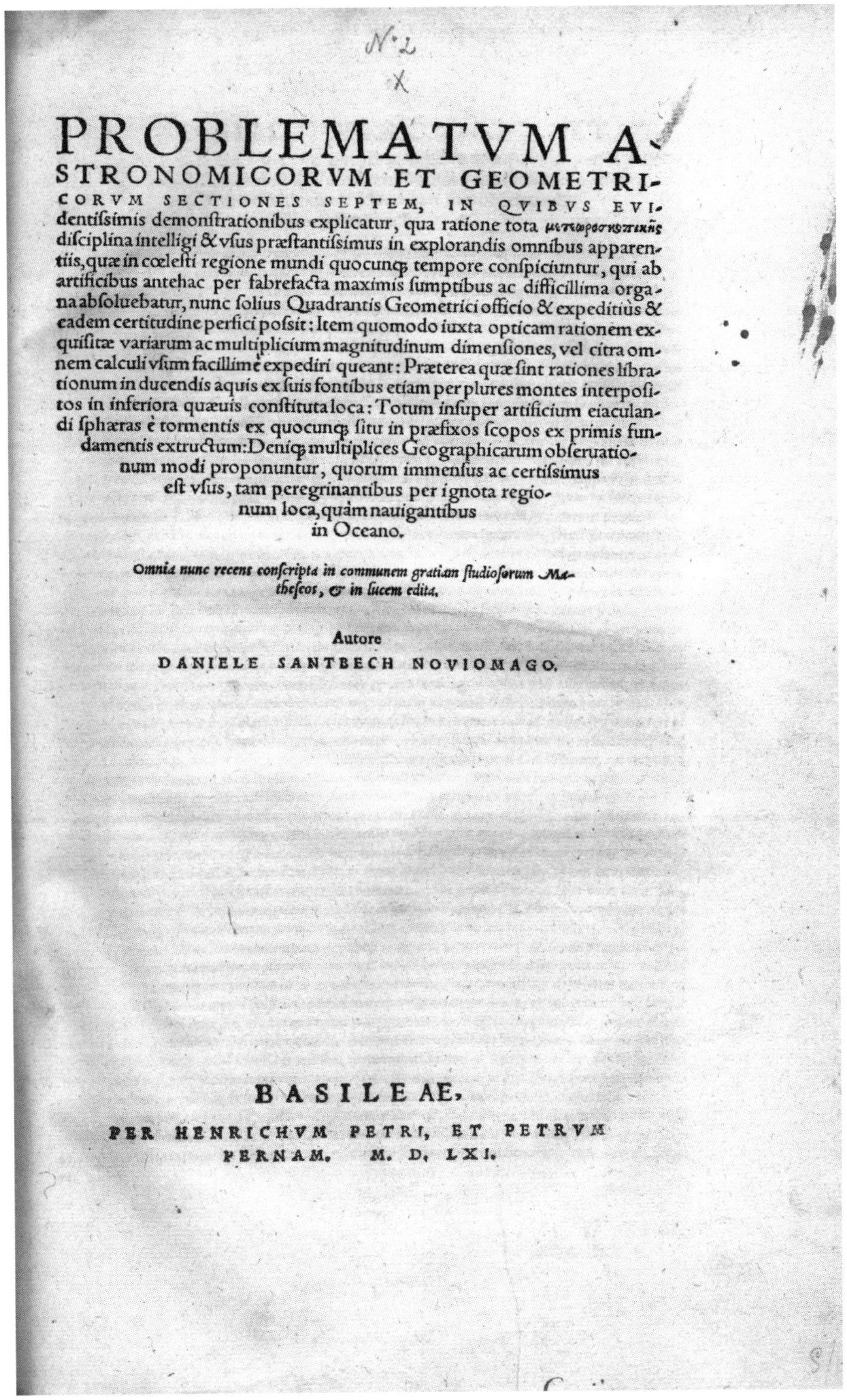
N° 2

PROBLEMATVM ASTRONOMICORVM ET GEOMETRICORVM SECTIONES SEPTEM, IN QVIBVS EVIdentissimis demonstrationibus explicatur, qua ratione tota μετεωροσκοπικῆς disciplina intelligi & vsus præstantissimus in explorandis omnibus apparentiis, quæ in cœlesti regione mundi quocunque tempore conspiciuntur, qui ab artificibus antehac per fabrefacta maximis sumptibus ac difficillima organa absoluebatur, nunc solius Quadrantis Geometrici officio & expeditiùs & eadem certitudine perfici possit: Item quomodo iuxta opticam rationem exquisitæ variarum ac multiplicium magnitudinum dimensiones, vel citra omnem calculi vsum facillimè expediri queant: Præterea quæ sint rationes librationum in ducendis aquis ex suis fontibus etiam per plures montes interpositos in inferiora quæuis constituta loca: Totum insuper artificium eiaculandi sphæras è tormentis ex quocunque situ in præfixos scopos ex primis fundamentis extructum: Denique multiplices Geographicarum obseruationum modi proponuntur, quorum immensus ac certissimus est vsus, tam peregrinantibus per ignota regionum loca, quàm nauigantibus in Oceano.

Omnia nunc recens conscripta in communem gratiam studiosorum Matheseos, & in lucem edita.

Autore
DANIELE SANTBECH NOVIOMAGO.

BASILEAE,
PER HENRICHVM PETRI, ET PETRVM PERNAM. M. D. LXI.

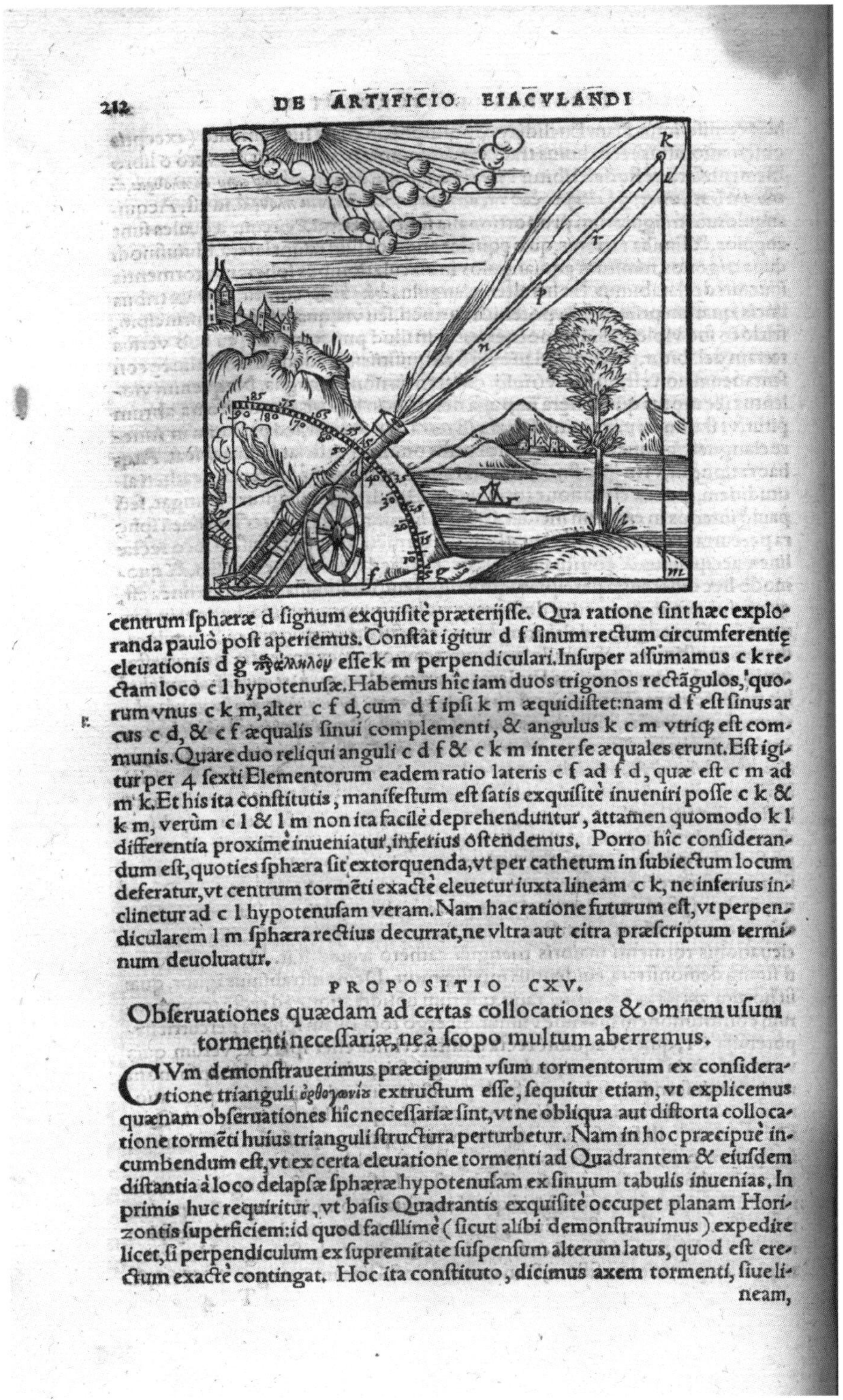

centrum sphæræ d signum exquisitè præterijsse. Qua ratione sint hæc exploranda paulò post aperiemus. Constat igitur d f sinum rectum circumferentiæ eleuationis d g παράλληλον esse k m perpendiculari. Insuper assumamus c k rectam loco c l hypotenusæ. Habemus hîc iam duos trigonos rectãgulos, quorum vnus c k m, alter c f d, cum d f ipsi k m æquidistet: nam d f est sinus arcus c d, & c f æqualis sinui complementi, & angulus k c m vtrique est communis. Quare duo reliqui anguli c d f & c k m inter se æquales erunt. Est igitur per 4 sexti Elementorum eadem ratio lateris c f ad f d, quæ est c m ad m k. Et his ita constitutis, manifestum est satis exquisitè inueniri posse c k & k m, verùm c l & l m non ita facilè deprehenduntur, attamen quomodo k l differentia proximè inueniatur, inferius ostendemus. Porro hîc considerandum est, quoties sphæra sit extorquenda, vt per cathetum in subiectum locum deferatur, vt centrum torm̃eti exactè eleuetur iuxta lineam c k, ne inferius inclinetur ad c l hypotenusam veram. Nam hac ratione futurum est, vt perpendicularem l m sphæra rectius decurrat, ne vltra aut citra præscriptum terminum deuoluatur.

PROPOSITIO CXV.

Obseruationes quædam ad certas collocationes & omnem usum tormenti necessariæ, ne à scopo multum aberremus.

CVm demonstrauerimus præcipuum vsum tormentorum ex consideratione trianguli ὀρθογωνίου extructum esse, sequitur etiam, vt explicemus quænam obseruationes hîc necessariæ sint, vt ne obliqua aut distorta collocatione torm̃eti huius trianguli structura perturbetur. Nam in hoc præcipuè incumbendum est, vt ex certa eleuatione tormenti ad Quadrantem & eiusdem distantia à loco delapsæ sphæræ hypotenusam ex sinuum tabulis inuenias. In primis huc requiritur, vt basis Quadrantis exquisitè occupet planam Horizontis superficiem: id quod facillimè (sicut alibi demonstrauimus) expedire licet, si perpendiculum ex supremitate suspensum alterum latus, quod est erectum exactè contingat. Hoc ita constituto, dicimus axem tormenti, siue lineam,

Fig. 53
German gunner's level and sight.
Catalogue no. 38.

38. Gunner's Level and Sight
Signed: I. D. F. 1621

While texts provided tables or calculating procedures to determine the range of shot as a function of the gun's elevation, some instrument makers offered similar information. This unusual instrument gives direct readings of distance for use with mortars, with scales for different kinds of shot.

The instrument consists of a foot with a vertical arc and a hinged upper part which carries a sighting tube. The inclination of the upper part is controlled by a screw. When the upper part is raised by turning the screw a pointer moves over the scales on the vertical arc. On one side the pointer is marked for 'fire bombs' ('FEVER K') and on the other for stone shot ('STEIN K'). By an ingenious system of transversal divisions the pointer is able to read each scale down to 10 unit intervals.

In use the instrument would have been longitudinally breech-mounted. It may have been placed with the vertical arc facing the breech and the screw pointing forwards. With the gun horizontal, the target would be sighted through the viewing tube (the direction of which should preferably be reversed from its orientation in figure 53). After estimating or measuring the distance to the target the pointer would be elevated to that distance on the scale, leaving the sighting tube to point downwards. The gun would then be elevated until the target again came in view through the tube.

In addition to acting as a level giving direct readings of range, the instrument probably also served as a more standard sight when the upper part of the instrument was set to zero elevation. The central upright which supports the sighting tube has two pillars on one of which there is a scale from 1 to 10. On the other pillar there is a groove of the same length, suggesting that the upright would once have contained a sliding sight.

Height: 86 mm
Inventory no. 50,613

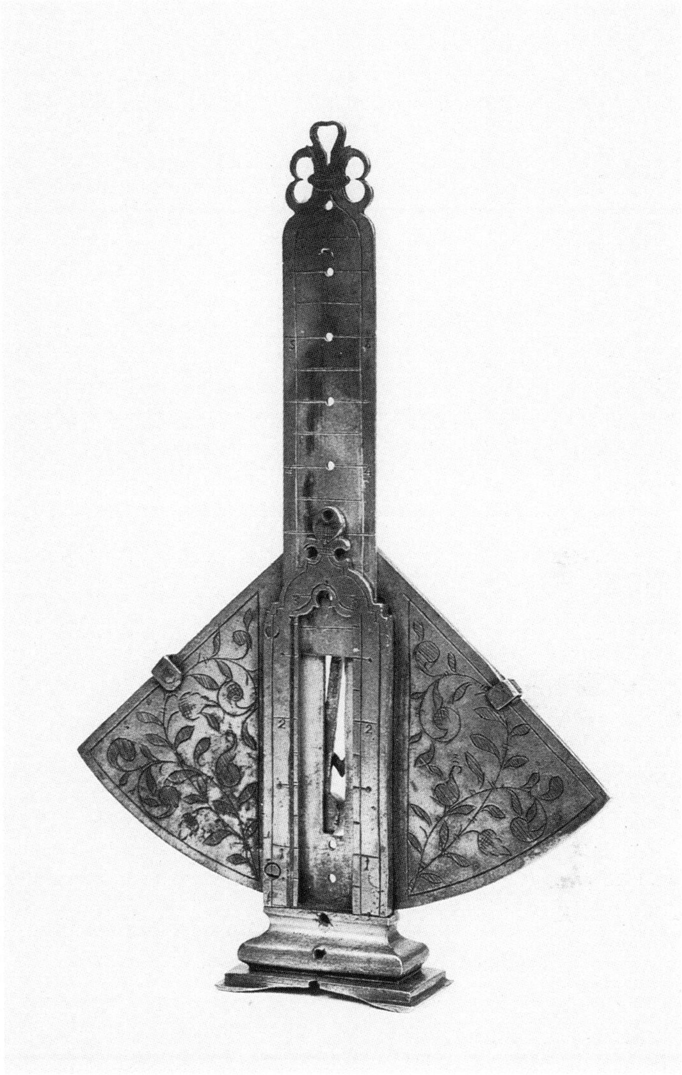

Figs 54 & 55
German gunner's level and sight.
Catalogue no. 39.

39. Gunner's Level and Sight
17th century

Although differing in its form from the previous instrument, this combined level and sight also provides direct indications of distance rather than elevation in degrees or points. While its levelling arc looks like many others, its scales are again marked for 'fire bombs' and stone shot. The scales start from opposite ends of the levelling arc and are engraved '2000 STEIN WVRF' (stone throw) and 'FEVER WVRF 1500' (fire throw) for their furthest ranges. The division of the scales has a geometric rather than empirical basis, being calculated as a tangent function (cf. Wunderlich, p. 186). The silvered brass plate for the levelling arc is slightly larger than a quadrant (about 100°) and decorated with foliage and birds.

The sight consists of a sliding plate pierced with holes, moving against a scale of 3 inches divided to quarters. When extended, the sliding plate reveals an extra scale of inches from 4 to 6 so that elevations above 3 inches can be set. The plummet has a central slot so that when the instrument is level the sighting holes are not obscured.

Height (slider not extended): 116 mm
Inventory no. 49,135

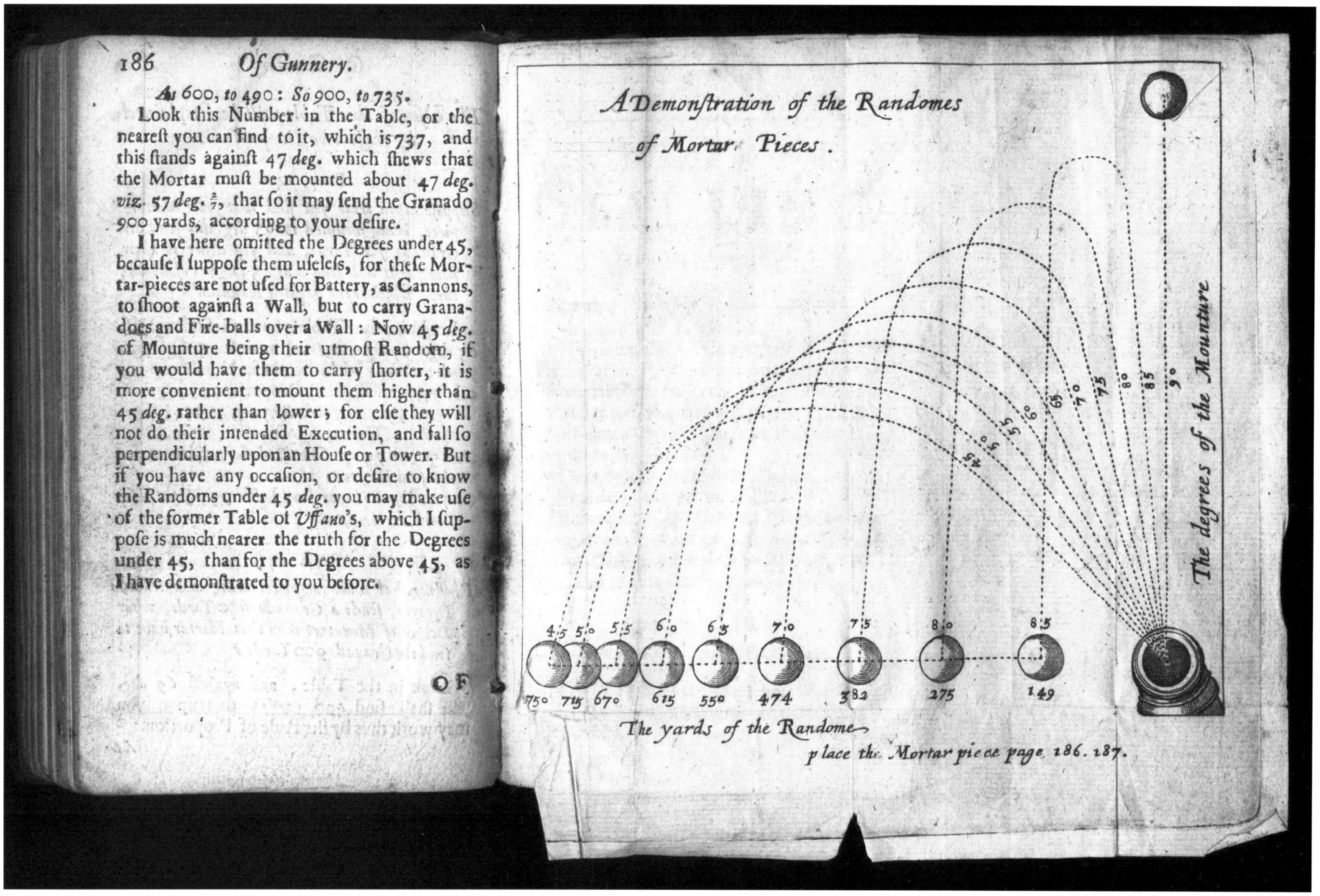

Fig. 56
Folding plate of
mortar trajectories.
From catalogue no. 40.

40. Henry Phillippes, *A Mathematical Manual* (first edition, 1669; this edition, London, 1693).

Phillippes was best known as an author on navigation and an editor of the popular English almanacs initiated by John Tapp. This text, providing tables of logarithms in a convenient portable format along with examples of their use, was intended by its author ‘only for a Manual or Pocket-Book’. Nevertheless, gunnery was sufficiently recognized as a component of English mathematical practice to come within the bounds of Phillippes’ treatment.

The folding plate illustrated in figure 56 – ‘A Demonstration of the Randomes of Mortar Pieces’ – shows the trajectories of mortar bombs from 45 to 90 degrees, at 5 degree intervals. The distances are based on (though not identical with) the values given by an earlier English mathematical practitioner, Robert Norton, who published several works on artillery including *The Gunner* of 1628. Phillippes published his graphical ‘demonstration’ of trajectories to render this aspect of artillery ‘more plain and visible’ and remarked that ‘though it differs much from other Mens Schemes thereof, yet I doubt not but that you will find it much nearer the truth, especially for the Ranges above 45, which are chiefly necessary’.

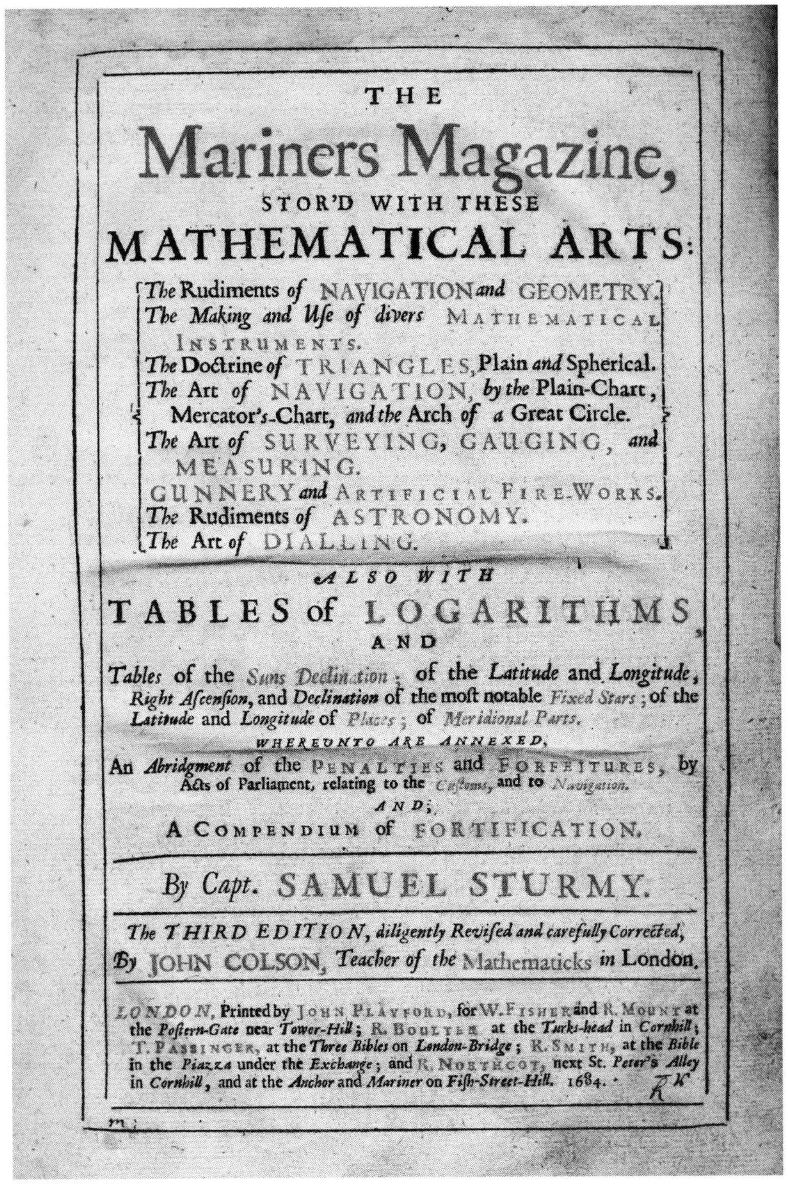

THE

Mariners Magazine,

STOR'D WITH THESE

MATHEMATICAL ARTS:

The Rudiments of NAVIGATION and GEOMETRY.
The Making and Use of divers MATHEMATICAL INSTRUMENTS.
The Doctrine of TRIANGLES, Plain and Spherical.
The Art of NAVIGATION, by the Plain-Chart, Mercator's-Chart, and the Arch of a Great Circle.
The Art of SURVEYING, GAUGING, and MEASURING.
GUNNERY and ARTIFICIAL FIRE-WORKS.
The Rudiments of ASTRONOMY.
The Art of DIALLING.

ALSO WITH

TABLES of LOGARITHMS

AND

Tables of the Suns Declination; of the Latitude and Longitude, Right Ascension, and Declination of the most notable Fixed Stars; of the Latitude and Longitude of Places; of Meridional Parts.

WHEREUNTO ARE ANNEXED,

An Abridgment of the PENALTIES and FORFEITURES, by Acts of Parliament, relating to the Customs, and to Navigation.

AND,

A COMPENDIUM of FORTIFICATION.

By Capt. SAMUEL STURMY.

The THIRD EDITION, diligently Revised and carefully Corrected, By JOHN COLSON, Teacher of the Mathematicks in London.

LONDON, Printed by JOHN PLAYFORD, for W. FISHER and R. MOUNT at the Postern-Gate near Tower-Hill; R. BOULTER at the Turks-head in Cornhill; T. PASSINGER, at the Three Bibles on London-Bridge; R. SMITH, at the Bible in the Piazza under the Exchange; and R. NORTHCOT, next St. Peter's Alley in Cornhill, and at the Anchor and Mariner on Fish-Street-Hill. 1684.

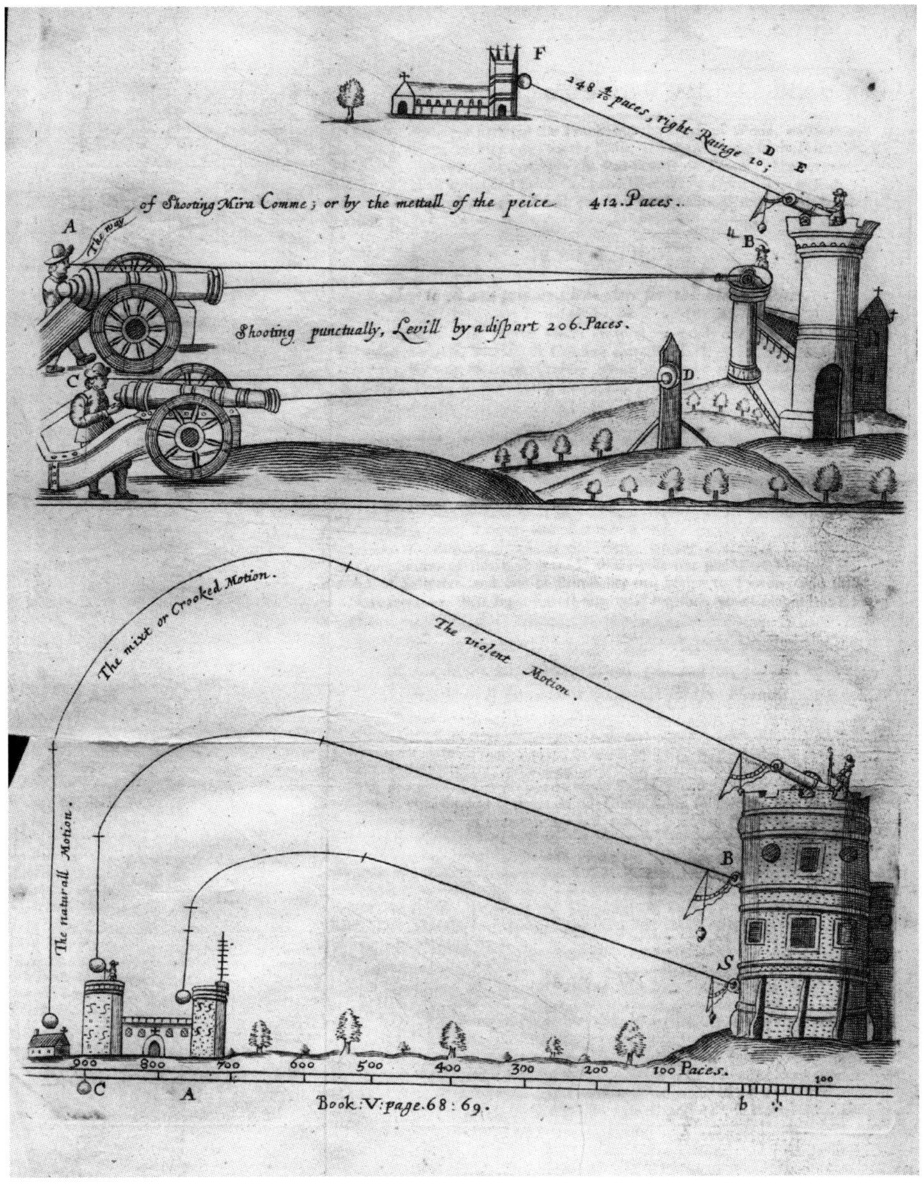

Figs 57 & 58
Samuel Sturmy's *Mariners Magazine:* title page and artillery plate.
From catalogue no. 41.

41. Samuel Sturmy, *The Mariners Magazine, or Sturmy's Mathematicall and Practicall Arts* (first edition, 1669; this edition, London, 1684).

Although identified as a work for mariners, Sturmy's text was more wide-ranging in its mathematical coverage. Aside from navigation and trigonometry, the title page listed its contents as surveying, gauging, astronomy and dialling. Gunnery and 'artificial fireworks' were also included and the addition of Staynred's tract on fortification (catalogue no. 75) brought the book closer to a comprehensive survey. Sturmy nevertheless gave a nautical slant to his treatment of gunnery when he commented in the preface: 'I am ashamed to hear how senselesly many Sea Gunners will talk of the Art, and know little or nothing therein'.

One of Sturmy's remedies for the defects of contemporary gunners lay in the provision of instruments and he both illustrated and discussed an elaborate gunner's scale. This scale had a stepped edge to provide a gauge for the bore of guns. In use, the scale was inserted as far as possible into the barrel of a gun; the step that did not enter indicated the name of the artillery piece being measured. Each type of gun had an entry in a table giving numerical information on the gun, its powder, ladle and shot, as well as its point blank range. This scale, which was to be made of 'Silver, Brass, or Box, or any other fine-grained Wood, that will not warp', also carried a quadrant for elevating artillery.

Sturmy was concerned more with the practical procedures which a gunner had to master than with the mathematical analysis of ballistics. Just like Tartaglia almost 150 years earlier, he divided the path of artillery projectiles into three parts: an initial straight line of violent motion, then a curved section of 'mixt or crooked motion' and finally the vertical descent to the ground of natural motion.

Figs 59 & 60
Gunner's rule by John Rowley.
Catalogue no. 42.

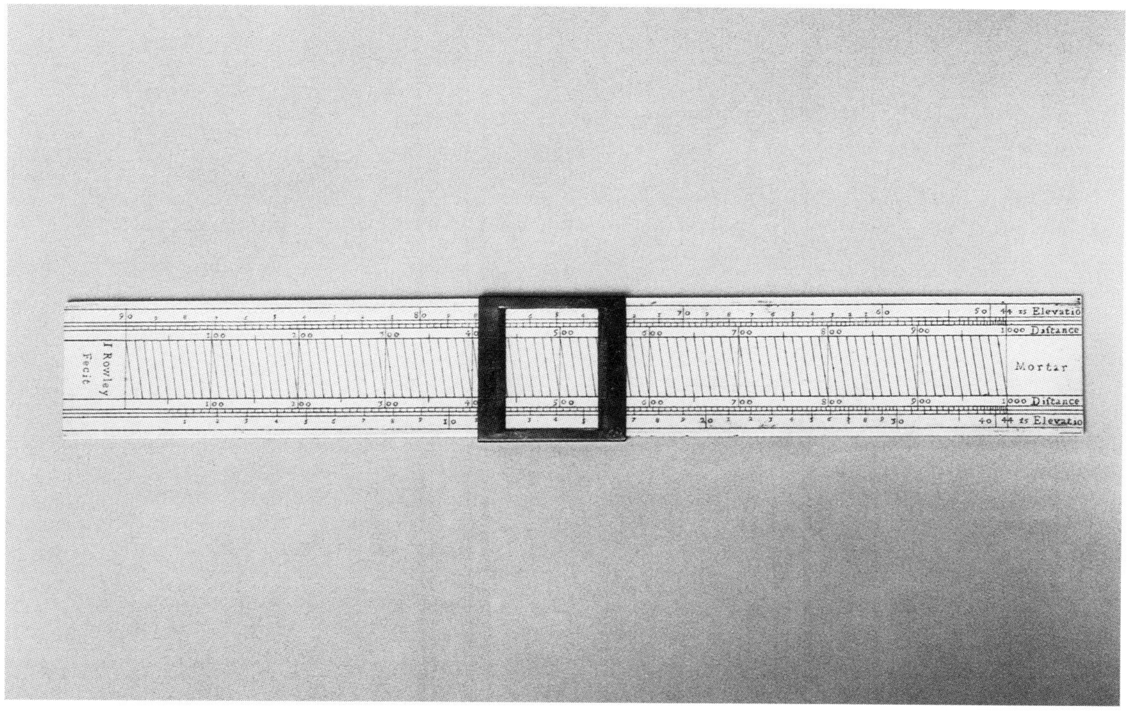

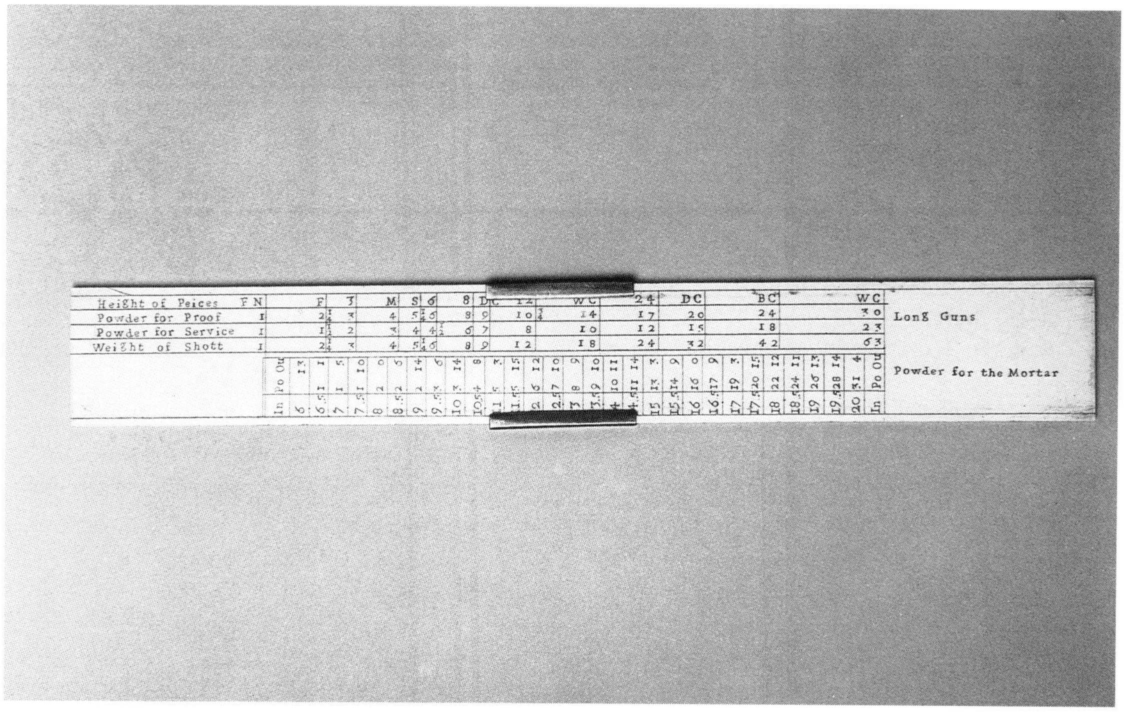

42. Gunner's Rule
Signed: I Rowley fecit
c. 1700

As mathematical instrument maker to the English Board of Ordnance in the early 18th century, John Rowley was accustomed to making instruments and gauges for artillery. However, this rule is not a standard-issue device. Its most distinctive feature is the diagonal scale labelled 'Mortar' which relates the range of a shot to elevation. The scale is based on a table of horizontal mortar distances by Robert Anderson, a mathematically-inclined silk-weaver of London, who published his *The Genuine Use and Effects of the Gunne* in 1674 (Appleby, p. 3). Anderson's table was reprinted by John Harris in his *Lexicon technicum* (first edition, 1704) under the heading 'Mortars' and a note was added:

> Since this, our Excellent Mathematical Instrument-maker, Mr. John Rowley, hath contrived this Table on a Scale of Box, where, by sliding only a fiducial Edge of Brass over the Diagonals of the Distance required, both the Elevations, upper and lower, are shewn at the same time.

The example shown here uses more impressive materials: ivory rather than box for the rule and silver rather than brass for the sliding cursor. In use, the rule provides a graphical solution to the problem of ranges. Given the range of a particular piece at a given elevation, a gunner could work by proportion to find the range at any other elevation and vice versa.

The reverse of the instrument carries more familiar material in the form of two tables for shot and powder. One gives the weight of shot and the amount of powder (for both proof and service) for artillery pieces; this provides simply an abbreviated version of the data on the Jobson folding rule (catalogue no. 3). The second table indicates the weight of powder required for mortars of different diameter; Rowley has again relied here for his information on the work of Robert Anderson.

Length: 249 mm
Orrery Collection (Christ Church)
Inventory no. 47,160

IN PLANISPH. LIB. IIII. 199

que accedere eos neque recedere, sed potius se à loco nõ mouisse colligemus.

Fig. 61
The use of the astrolabe in warfare.
From catalogue no. 43.

43. Juan de Rojas, *Commentariorum in astrolabium* (Paris, 1551).

Astronomical calculation, including related problems such as time-telling, is regarded as the principal function of the astrolabe and the ingenuity and sophistication of both maker and designer are particularly evident on the front of the instrument, where such calculations are performed. However tracts on the astrolabe often seek to illustrate its general usefulness, and the back is particularly helpful to authors intent on demonstrating a universal functionality. This is because the pivoted alidade or sighting rule moving across a circular scale of degrees, whose astronomical use would be for taking the altitude of a star or of the sun, has many possible applications. It can be used in surveying, for example, or in any enterprise involving the measurement of altitudes (vertical angles) or bearings (horizontal angles). The point is reinforced on the backs of most astrolabes by the inclusion of a 'shadow square' or 'geometrical quadrant' (see figure 62), where such angles can be measured as ratios (in trigonometrical terms, as tangents expressed as fractions). In the case of an altitude measurement, for example, the ratio would be that between the horizontal and vertical distances from the observer to his target; if either of these distances are known, the other can readily be calculated using the measured ratio.

Military applications are often cited in accounts of the usefulness of the back of the astrolabe. Its general relevance to surveying includes military installations and fortifications, and finding the altitude of the target is necessary for setting a gun. It can also be applied to range-finding. This could be done by triangulation with the instrument horizontal. Alternatively, the shadow square in a vertical orientation can measure the ratio between an object whose approximate height is known, or easily estimated, and the horizontal distance required by the gunner.

The tract selected here is by Juan de Rojas, whose name is associated with a type of universal astrolabe: it carries a particular projection of the spherical co-ordinate lines of the heavens and can be used for astronomical calculation in any latitude. However, despite the ingenuity of this development on the front of the instrument, a significant proportion of his book is devoted, as was customary, to the application of the scales on the back to surveying in civil and military contexts.

Rojas's book is dedicated to his patron, Emperor Charles V, who had a sustained interest in practical geometry and mathematical instruments. The example Rojas chooses to illustrate the relevance of the astrolabe to warfare is the celebrated expedition against Tunis, led in person by Charles in 1535. The engraving showing the astrolabe in use suspended from a lance (*hasta*) may well be intended to depict the Emperor himself (see figure 61).

44. Astrolabe

Signed: ANTHOITNE MESTREL APARIS 1551

This instrument was made in Paris in 1551 and is closely related to the book on the astrolabe by Rojas (catalogue no. 43), which was published in Paris in the same year. There is, for example, a strong resemblance between the illustrations of the astrolabe in the book and the instrument itself. One side of the astrolabe has the Rojas universal projection, while the other – the back – has a pivoted alidade with a shadow square and degree scale, a zodiacal calendar and a diagram for converting times between different systems of hours (known as equal and unequal). It is the alidade and shadow square on the back, and possibly the scale of degrees, that were said to have applications to gunnery and military surveying.

Diameter: 257 mm
Billmeir Collection
Inventory no. 32,378

45. Leonard and Thomas Digges*, A Geometrical Practise Named Pantometria* (first edition, 1571; this edition, London, 1591).

Features of the back of the astrolabe could be transferred to dedicated instruments for surveying, leading to the design of a number of instruments where a pivoted alidade moves over a plate engraved with a circular scale of degrees, a geometrical quadrant (shadow square), or both. Such an instrument, like the astrolabe itself, would take angles in a single plane – vertical or horizontal. If the geometrical quadrant dominated the design, the instrument would be so called; if the circular degree scale were more pronounced, a variety of names were used but 'theodolite' became the most common. A theodolite of this sort – a 'simple theodolite' – differed from the modern instrument in having only one divided circle.

Leonard Digges was one of those who proposed a more ambitious configuration and in the book *Pantometria*, edited and completed by his son Thomas, he arranged the scales of the astrolabe into his 'topographical instrument', for taking vertical and horizontal angles together. The horizontal plate was a divided square with an inscribed

Fig. 62
French astrolabe, 1551.
Catalogue no. 44.

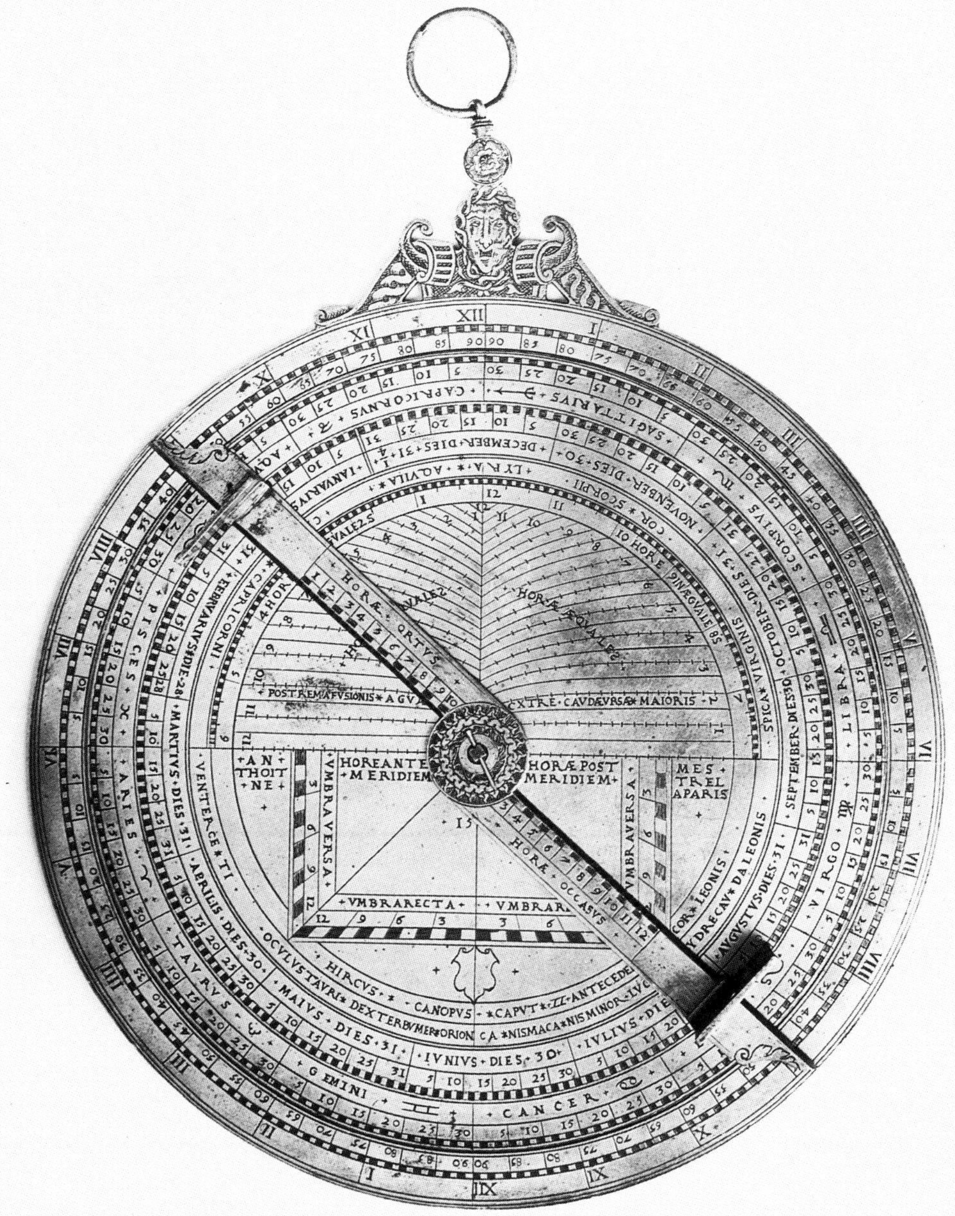

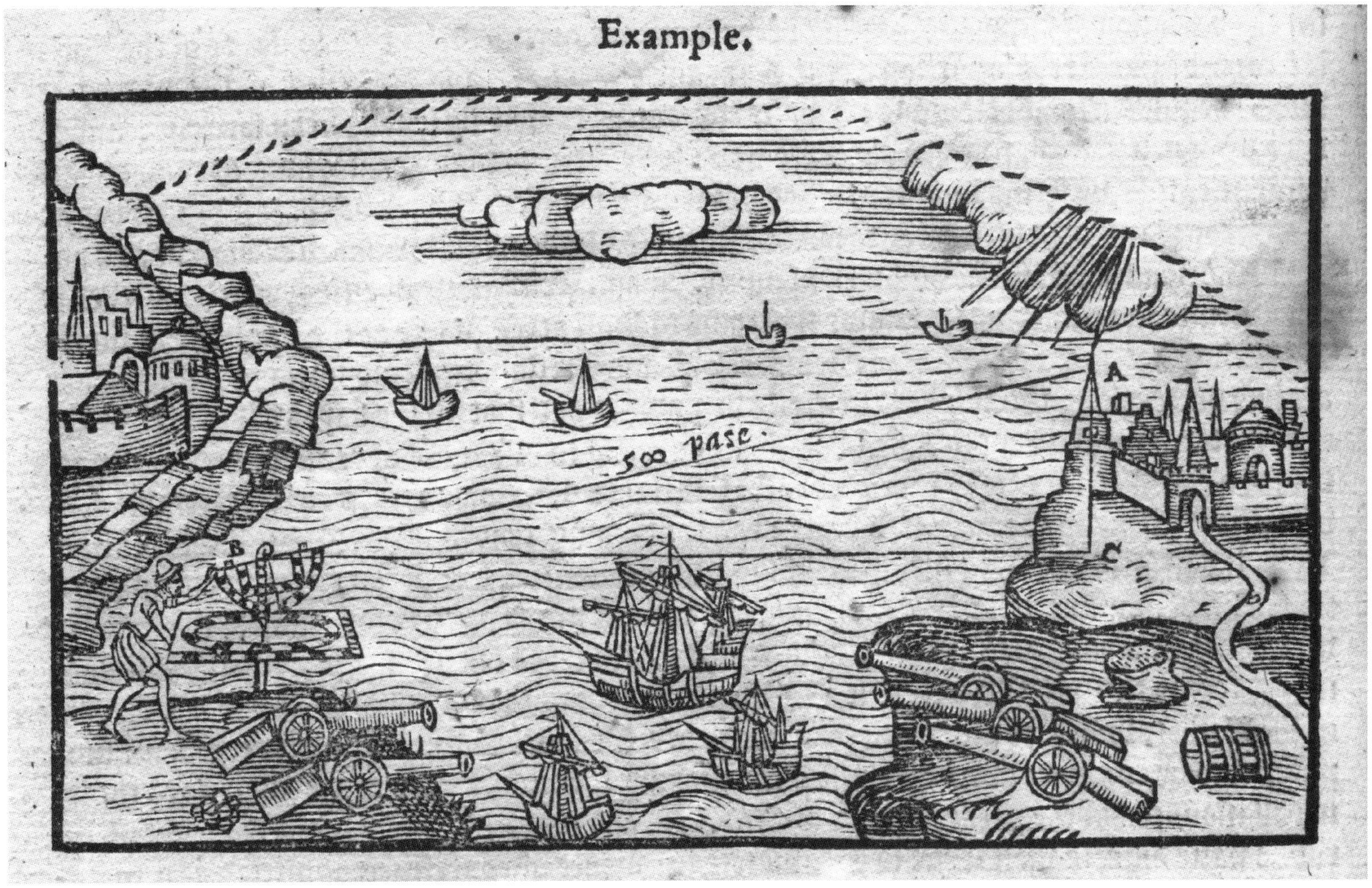

Fig. 63
Range-finding with Digges's 'topographical instrument'.
From catalogue no. 45.

circle, while the vertical was a semicircle with an inscribed double shadow square. By this combination altitudes or bearings could be measured as ratios or as degrees (see catalogue no. 43) and the result was a universal instrument, 'seruing most commodiously for all manner mensurations', not least those required for war.

Digges explains how, to get the range of a ship passing a cliff or other vantage point, the gunner should note the reading from the vertical shadow square and, as he puts it, 'then measure the height of the cliffe, or rather the centre of your semicircles altitude aboue the sea'. While he was confident of the usefulness of this technique for range-finding, he pointed out that setting the ordnance for the measured range required a further area of proficiency, but this too was amenable to geometry, or rather to the practical geometer's characteristic combination of mathematical reason and experience.

> This conclusion serueth most commodiouslie for all such as shall haue committed to their charge any platforme with ordinaunce, for hereby you may exactly at the first view, tell the distance of any ship or marke, so that hauing a table of Randons made, mounting your peeces accordingly, no vessell can passe by your platforme … but you may with your ordinance at the first bouge hit and neuer bestowe vaine shotte. … But to giue full instruction for shooting great ordinance, and for the variation of Randons … would require a long discourse: I haue thereof in a booke by itself very largely, and I dare be bold to saye, sufficiently entreated: for conferring and conioyning Geometricall demonstration, with my long continued painefull practices, I haue at last reduced that most irregular course & circute of the Bullet (framed and compounded of violent and naturall motions) within the boundes of numbers and Arithmeticall rules, which whether it be difficulte or no, they onely knowe that haue or shall attempt to doe the like. This surely I will say, that as the ignorant in Geometricall and Arithmeticall proportions, shall neuer attaine perfection, though hee turmoile in powder and shot all the dayes of his life: so the Geometer, how excellent so euer he be, leaning onely to discourse of reason, without practise (yea and that sundrie wayes made) shall fall into manifolde errors, or inextricable Laberinthes.

Digges goes on to make it clear that he assigned Tartaglia (see catalogue no. 1) to the latter group; in England in 1571 the geometrical description of the flight of shot was already a matter for discussion and contrary opinion.

The title-page is missing from this copy.

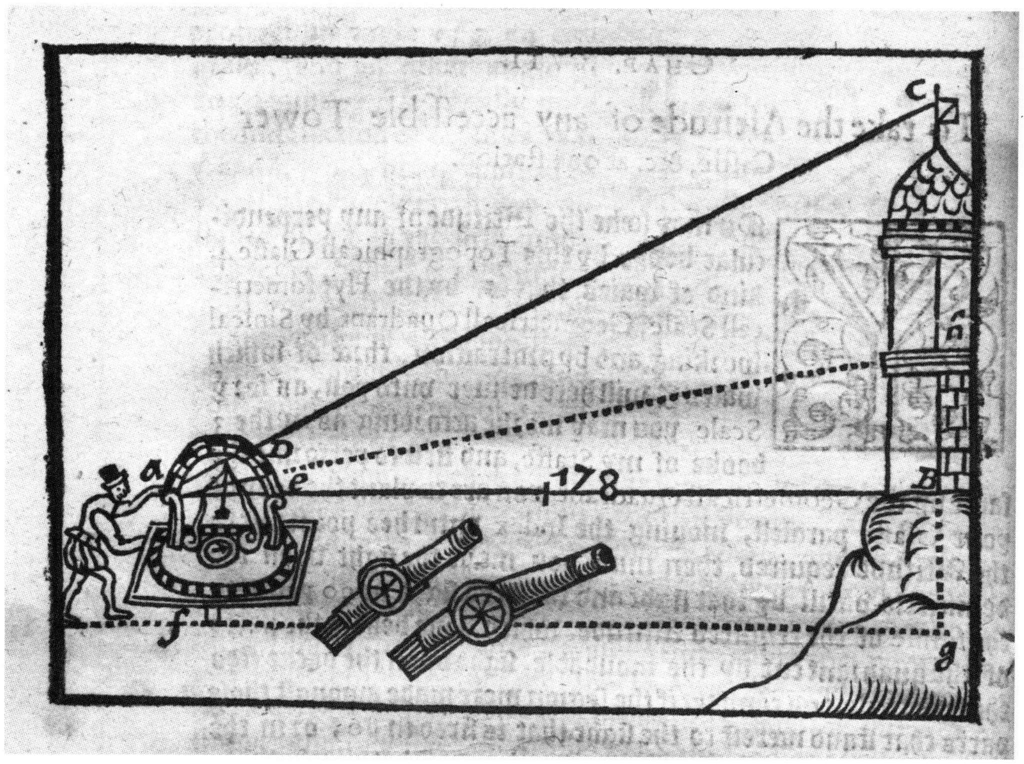

Fig. 64
Arthur Hopton's topographical glass applied to range-finding.
From catalogue no. 47.

46. Altazimuth Theodolite
Signed: H. Cole. 1586

Digges designed the horizontal plate of his 'topographical instrument' as a single geometrical quadrant with an inscribed circle divided by degrees. The instrument in the exhibition, by the man generally regarded as the first native English instrument maker, Humphrey Cole, has a vertical semicircle very similar to that described by Digges, which now sits above a horizontal circle with an inscribed square. Each quarter of the square has sides divided as a geometrical quadrant and this alternative arrangement has the advantage that a single alidade, pivoted at the centre, can be used with the circle or with any of the four geometrical quadrants. By the same token, the alidade fixed to the vertical semicircle can be used for measurements of both coordinates. The arrangement of scales in the Cole instrument is in fact closer to an earlier design of altazimuth instrument, the 'polimetrum' of Martin Waldseemüller

The confusion of terminology, embracing both 'shadow-square' and 'geometrical quadrant' for the same types of scale, is illustrated on this instrument, where the vertical scales are marked 'Vmbra Versa' and 'Vmbra Recta', and the horizontal 'Quadratum Geometricum'. Certainly the use of such a scale to convert a measurement of a shadow cast on the ground into a height, was an application of the vertical scale.

Despite the refinements to the Digges design, the Cole altazimuth theodolite is conceptually similar to the topographical instrument, which promised that the gunner would 'neuer bestowe vaine shotte'.

Diameter (horizontal circle): 197 mm
St John's College Collection
Inventory no. 55,130

47. Arthur Hopton, *Speculum topographicum: or The Topographicall Glasse* (London, 1611).

Hopton's 'topographical glass' was a form of altazimuth theodolite, but the division and configuration of the vertical circle, the mounting of the alidade and the scales on the horizontal plate were all different from Digges's topographical instrument (catalogue nos 45 and 46). Hopton was fully aware of the relationship between his design and that of Digges and, before giving an account of its use, felt it necessary to explain the advantages of his 'glass' over the earlier instrument. The examples he then offers of its applications include range-finding and military surveying.

One of the cited advantages of the topographical glass was the space in the horizontal plate for a larger central compass. The compass needle was pivoted over a printed card with a degree scale and, in addition, a form of sundial making use of the relation between solar declination, solar azimuth, and time at a particular latitude. When using the compass and degree scale the instrument, as Hopton says, acts as a circumferentor. He offers the reader assistance in supplying this part of his instrument: '… and because I will not trouble you ouerlong with the making, behold the figure which I will cause to be printed in a voyd paper to saue labour in drawing the same'. The woodcut of the compass card is indeed reproduced on an extra page at the end of the volume, which also carries advertisements for the London mathematical instrument makers John Thompson and Elias Allen.

Fig. 65
Altazimuth theodolite
by Humphrey Cole,
1586.
Catalogue no. 46.

Fig. 66
Irish circumferentor,
1667.
Catalogue no. 48.

48. Circumferentor

Signed: W. R. Dublin. fe. Latitude 53 20 * 1667 *

A very early example of a circumferentor, perhaps the earliest known, that is also unusual in being Irish and in being made largely of wood. Wooden instruments were probably fairly common at the time, but have not survived well, and circumferentors were certainly more commonly used than something so complicated as an altazimuth theodolite. This rare survival is probably much more typical of useful, workaday instruments than the record of our collections would suggest.

The printed compass card follows Hopton's suggestion in combining a degree scale with a central sundial (see catalogue no. 47). He explained that a circumferentor could be used 'to plant barrels of powder, direct vnder Castles, Forts or such like'. Indeed the instrument became generally associated with mining, since the magnetic needle could be used for orientation underground, where no other sights were possible.

In a poetic preface to the *Topographicall Glasse*, Hopton includes among his potential readers and users those who:

Wouldst plant a Towne, or situate a house,
Or force the water take a wished course.
Wouldst seeke the heigth or distance of a Towne,
Or vndermine a Fort to blow it downe.

Length: 350 mm
Inventory no. 33,340

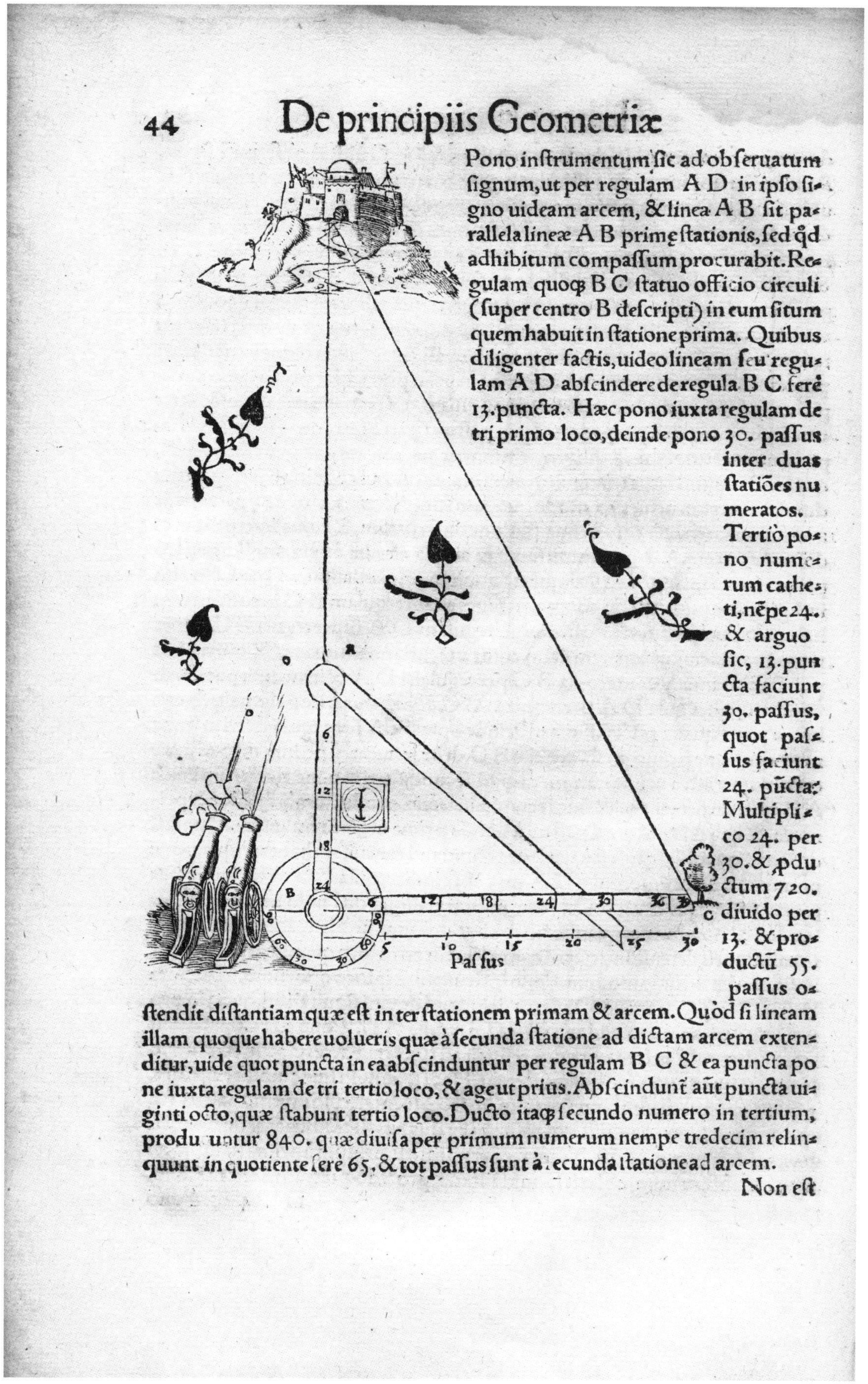

44 De principiis Geometriæ

Pono instrumentum sic ad obseruatum signum, ut per regulam A D in ipso signo uideam arcem, & linea A B sit parallela lineæ A B primę stationis, sed q̄d adhibitum compassum procurabit. Regulam quoque B C statuo officio circuli (super centro B descripti) in eum situm quem habuit in statione prima. Quibus diligenter factis, uideo lineam seu regulam A D abscindere de regula B C ferè 13. puncta. Hæc pono iuxta regulam de tri primo loco, deinde pono 30. passus inter duas statiões numeratos. Tertiò pono numerum catheti, nēpe 24. & arguo sic, 13. puncta faciunt 30. passus, quot passus faciunt 24. pūcta? Multiplico 24. per 30. & productum 720. diuido per 13. & productū 55. passus ostendit distantiam quæ est inter stationem primam & arcem. Quòd si lineam illam quoque habere uolueris quæ à secunda statione ad dictam arcem extenditur, uide quot puncta in ea abscinduntur per regulam B C & ea puncta pone iuxta regulam de tri tertio loco, & age ut prius. Abscindunt aūt puncta uiginti octo, quæ stabunt tertio loco. Ducto itaque secundo numero in tertium, producuntur 840. quæ diuisa per primum numerum nempe tredecim relinquunt in quotiente ferè 65. & tot passus sunt à secunda statione ad arcem.

Non est

Fig. 67
The triangulation instrument illustrated by Münster.
From catalogue no. 50.

49. Levinus Hulsius, *Instrumenta mechanica* (Frankfurt, 1605).

The *Tractatus primus* deals, along with other surveying instruments, with the traditional geometrical quadrant, and the folding plates illustrating its use seek to place surveying within the general context of waging war.

50. Sebastian Münster, *Rudimenta mathematica* (Basel, 1551).

Münster outlines the principles of geometry, as an introduction to a more substantial treatise on dialling, and includes range-finding applications in his account of triangulation by an instrument.

51. Benjamin Bramer, *Bericht zu M. Jobsten Burgi seligen geometrischen triangular Instruments* (Kassel, 1648).

Bramer describes a triangulation instrument whose design he attributes to the instrument maker and clockmaker Joost Bürgi, who was employed by Tycho Brahe. Books explaining the use of such devices are not uncommon in the period, even though the instruments themselves are rare. They are always presented as universal in their applications, among which military uses are given prominence.

In figure 68, Bürgi's instrument is shown used by gunners on a rampart overlooking an estuary, beyond which a company of foot soldiers advances on an undefended fort. A baseline is measured between the two positions of the instrument, the sliding pivot of one sighting arm is adjusted so that the distance between the pivots is a scaled representation of the baseline, and sights are taken from both positions. The sides of the triangle on the ground – similar to that formed by the instrument – are then found by applying the scale to the intersection readings of the pivoted arms.

Fig. 68 (right)
Bürgi's triangulation instrument solves an urgent problem for the gunners.
From catalogue no. 51.

Fig. 69 (below)
The geometrical quadrant applied to warfare.
From catalogue no. 49.

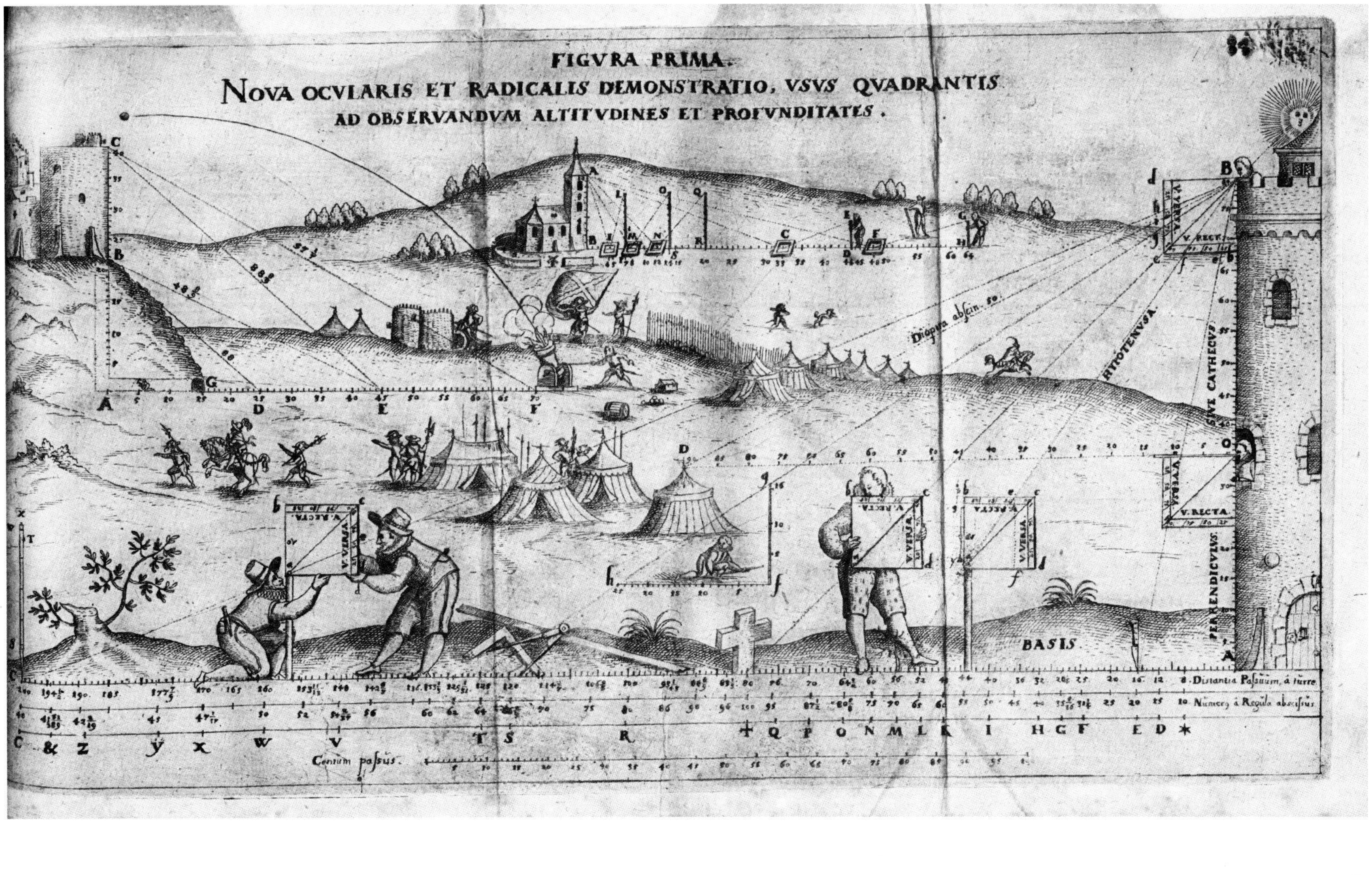

Geometrischen Instrument.

Fig. 70 (far left)
Triangulation instrument by Joost Bürgi.
Catalogue no. 52.

Fig. 71 (near left)
An example of Zubler's 'new geometrical instrument' made by Christoff Tressler, with the printed account by Zubler.
Catalogue no. 55.

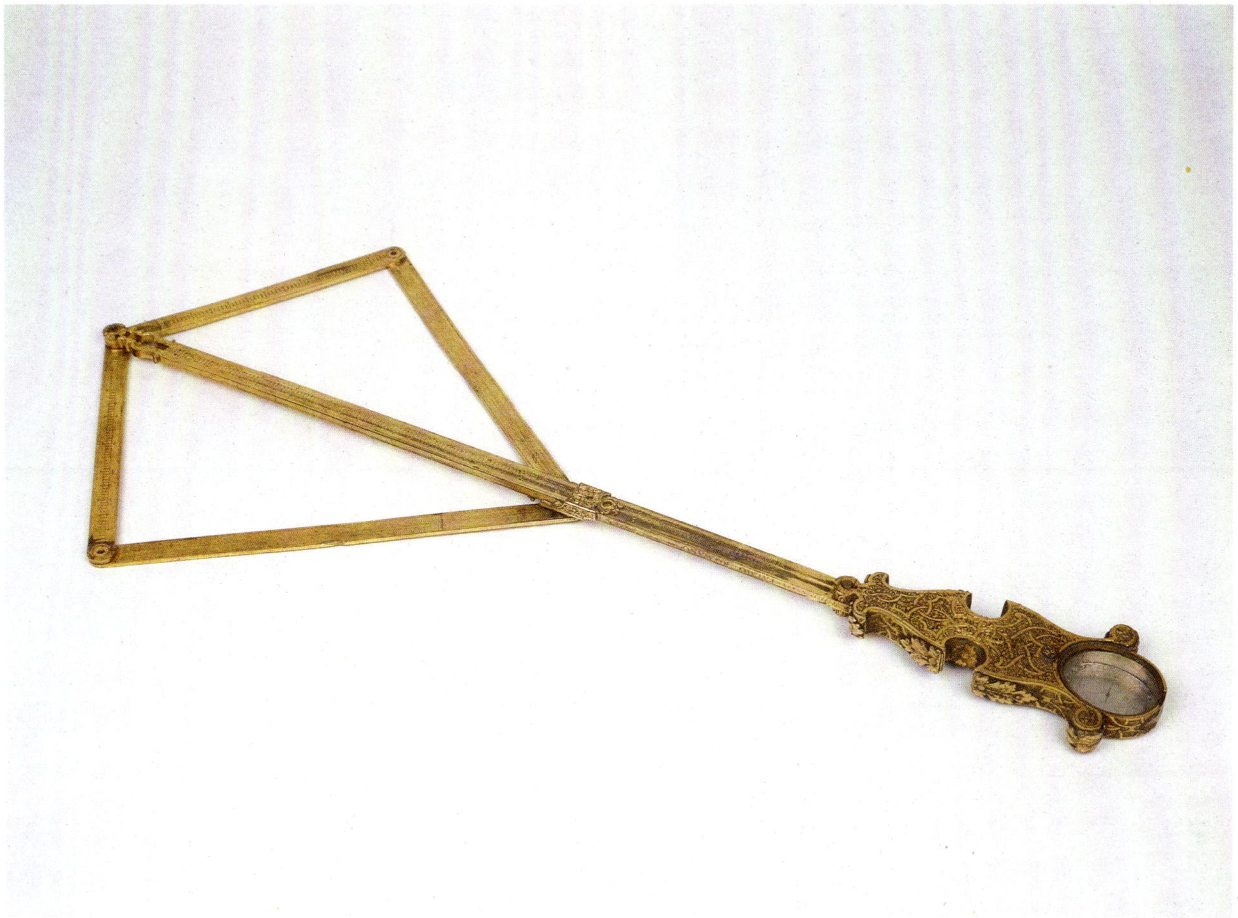

Fig. 73
The *radio latino*, unfolded.
Catalogue no. 60.

Fig. 74
Military surveying instrument with turned wooden case.
Catalogue no. 61.

Fig. 72 (left)
Triangulation instrument by Erasmus Habermel, with leather-covered case.
Catalogue no. 53.

52. Triangulation Instrument

Signed: Iost Bürgi F
c. 1600

Catalogue no. 52 is illustrated in colour on page 60.

A gilt brass triangulation instrument by the maker to whom the design is ascribed by Bramer. The instrument coincides with that illustrated in catalogue no. 51, having one fixed and one sliding pivot, and each of the three arms engraved with a linear scale. As if to reinforce the gunnery applications, the reverse side of one arm has scales relating the size of shot to weight for three materials, marked in German, 'BLEI', 'GROSSEN EISEN' and 'STEIN' (see catalogue no. 4). The sights are in steel and the whole is very finely made, though without decoration.

Length (baseline arm): 324 mm
Billmeir Collection
Inventory no. 54,181

Fig. 75 (top right)
The use of Zubler's triangulation instrument for range-finding.
From catalogue no. 54.

Fig. 76 (bottom right)
Letters engraved on the instrument by Tressler coincide with those on a plate in the published description by Zubler.
Catalogue nos 54 & 55.

53. Triangulation Instrument

Signed: Fecit Erasmus habermel
Late 16th century

Catalogue no. 53 is illustrated in colour on page 60.

Catalogue no. 55 is illustrated in colour on page 60.

Erasmus Habermel, like Joost Bürgi, enjoyed the patronage of Rudolf II at the imperial court in Prague, and it may well have been through contact with Bürgi (see catalogue nos 51 and 52) that he came to make an instrument of this type. Habermel's version, in gilt brass, is more substantial and elaborate, but fundamentally similar. He too includes scales for shot, but here marked in Latin: 'PLVMBVM', 'FERRVM' and 'LAPIS'. The instrument has a shaped and tooled leather case.

Length (baseline arm): 470 mm
Billmeir Collection
Inventory no. 47,741

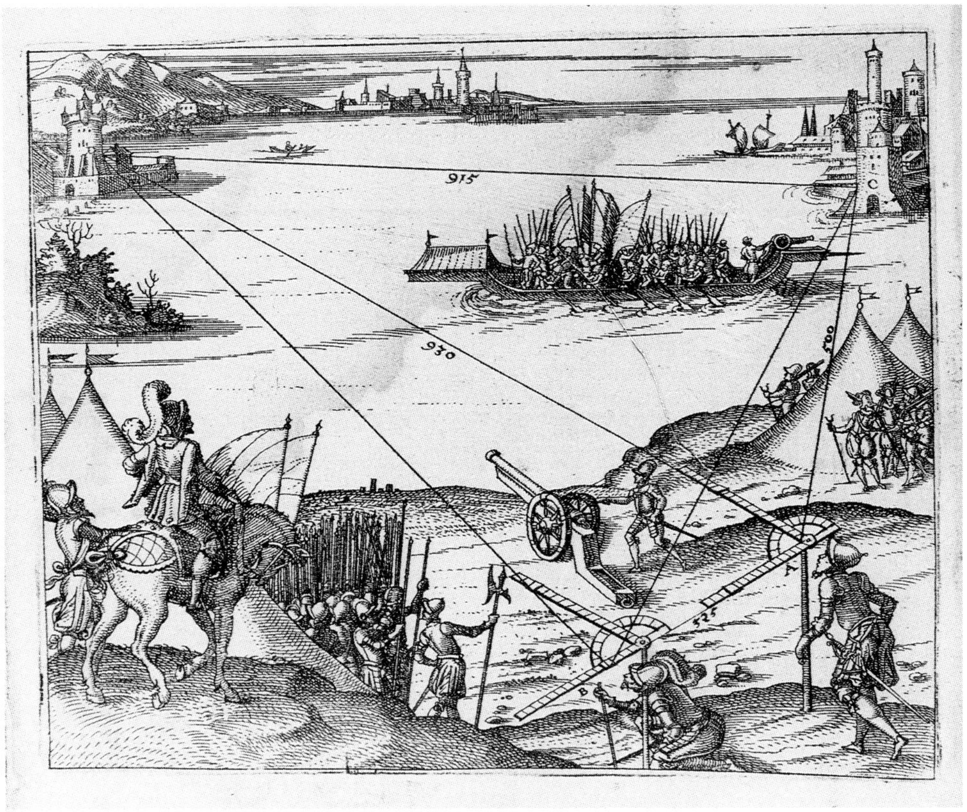

54. Leonhard Zubler, *Novum instrumentum geometricum* (Basel, 1607).

Zubler's book is another instance of the early-modern fashion for writing on somewhat exotic surveying instruments and emphasising military problems among their universal applications. Zubler's design is a further example of a triangulation instrument and its relevance to range-finding and to surveying fortifications is probably even more prominent that was usual in such books.

Zubler's triangulation instrument (see figure 71) dispenses with the third arm of the Bürgi design (figures 68 and 70), but can be used in the same way by fixing the angle between the baseline and the pivoted arm, setting the sliding sight on the baseline arm to the scaled length of the baseline itself and finally, from the second station, sighting across the two arms while moving the second sliding sight into alignment with the first sight and the target. The position of this second sight, adjusted to scale, then gives the side of the triangle on the ground. The provision of a degree scale on a semicircular plate at the pivot permits other techniques involving angle measurement or setting the arms at right angles.

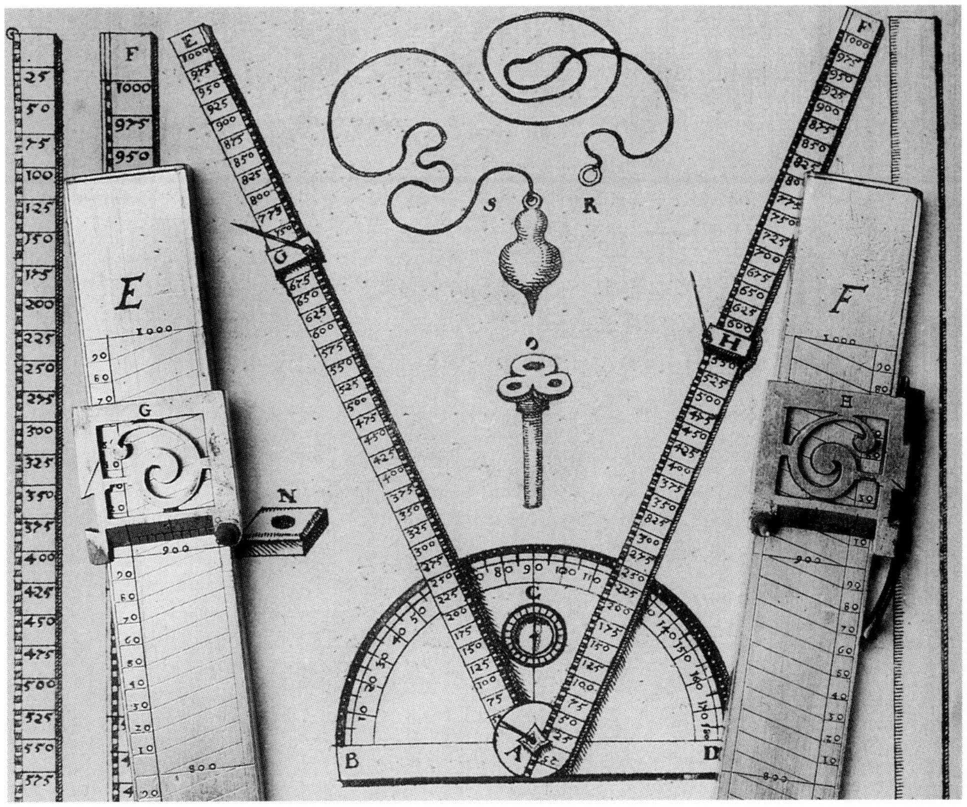

55. Triangulation Instrument

Signed: Christoff Tressler der Elter Mechanicus Anno 1617

This instrument reproduces Zubler's 'new geometrical instrument' (see catalogue no. 54), the only substantial differences from the illustrations in Zubler's book being the replacement of the central plate by a semicircular arc, and the absence of a compass. The arms carry linear scales, each graduated from 0 to 1000, as recommended by Zubler; diagonal lines are ruled across each division of ten, so that the scaled length can be read to 1 according to where one of these diagonals cuts a transverse scale, divided 1 to 10, on a bevelled edge of each sliding sight. The link to Zubler's book is very strongly reinforced by the fact that engraved letters on the instrument – 'A' at the pivot, 'B', 'C' and 'D' on the semicircle, 'E' and 'F' at the ends of the arms and 'G' and 'H' on the moving sights, coincide with the key letters on the first plate in the book, which illustrates the detailed construction of the instrument.

The reverse has a pair of sectoral scales, each marked: 'Linea recta divisio'.

Length (leg): 286 mm
Inventory no. 51,706

Fig. 77 (left)
Surveying instrument for fortification, 1579.
Catalogue no. 64.

Fig. 78 (right)
French surveying instrument and sundial with a scale of polygons, 1652.
Catalogue no. 63.

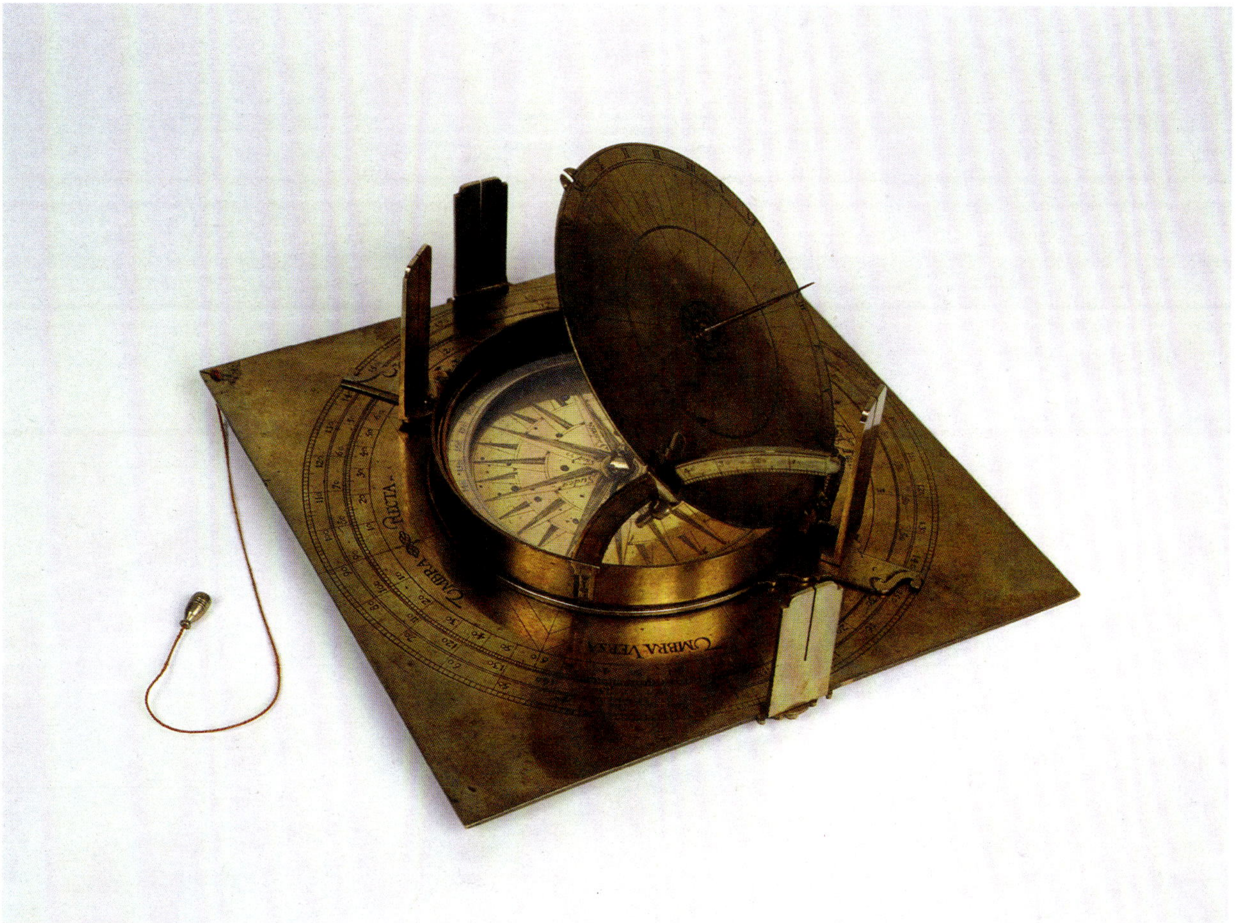

Fig. 79 (right)
Graphometer by Baradelle.
Catalogue no. 57.

Fig. 80 (below)
Military surveying with the graphometer.
From catalogue no. 56.

56. Philippe Danfrie, *Declaration de l'usage du graphometre* (Paris, 1597).

The instrument that Danfrie describes is similar to a simple theodolite (see catalogue no. 45), but with a semicircular scale replacing the full circle, and he describes its use for vertical as well as horizontal angles. (Though oddly, and presumably mistakenly, in the illustration opposite, the instrument is given a circular scale.)

Danfrie explains how to survey a fortified town from a distance, using a measured baseline. He points out that this can be done from as far away as the eye can manage and out of danger from cannon-fire.

57. Graphometer
Signed: Baradelle Æ Paris
Second half of the 18th century

The graphometer became popular among surveyors in France and this example, made almost two centuries after Danfrie's design, is from a Paris workshop of the later 18th century.

Length: 225 mm
Inventory no. 46,742

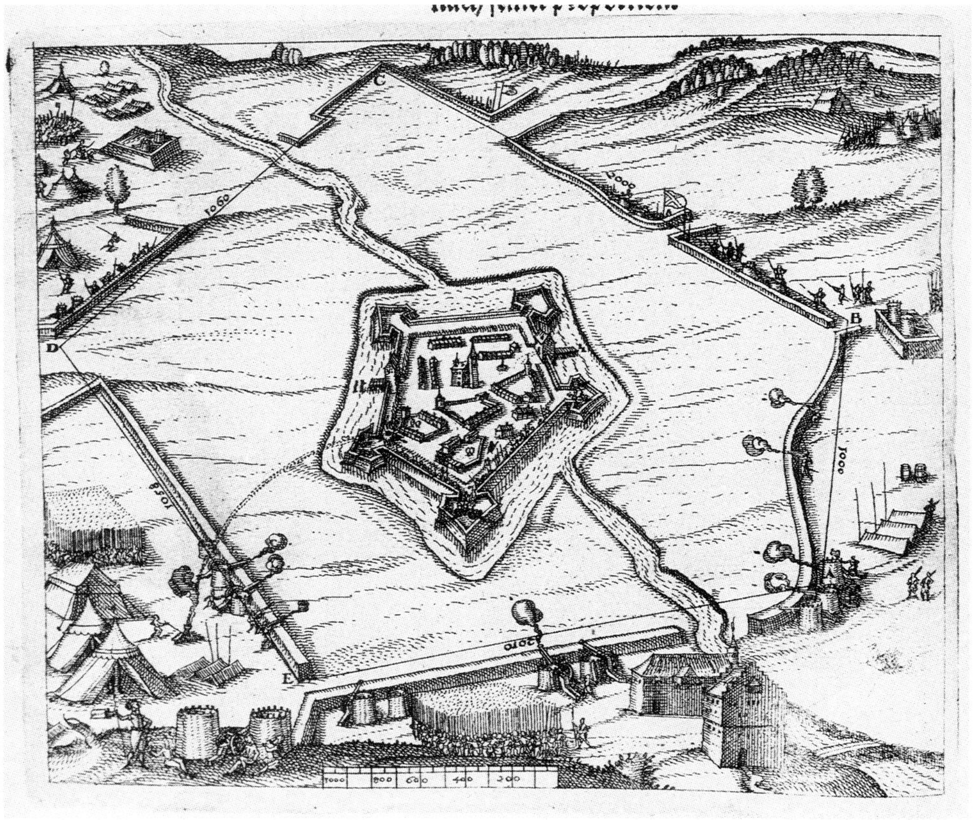

Fig. 81
Siege and bombardment of a fortified town.
From catalogue no. 58.

Fig. 82 (below)
An illustration explaining the relevance of the scales on the *radio latino* to fortification.
From catalogue no. 59.

58. Leonhard Zubler, *Fabrica et usus instrumenti chorographici* (Basel, 1607).

This tract by Zubler is mainly concerned with the use of the plane table, a versatile and popular instrument of surveying. As usual in his work, Zubler takes the opportunity to include a military scene (figure 81).

59. Egnatio Danti, Latino Orsini, *Trattato del radio latino* (Rome, 1583).

The basic text is by the instrument's designer, Orsini, augmented by extensive annotations by Danti, formerly cosmographer to Cosimo I de' Medici, but by the time of this publication in the service of Pope Gregory XIII. This example was a presentation copy from Danti and it has been in the library of the Collegio Romano.

The *radio latino* has five jointed arms engraved with scales, which can be arranged in different configurations depending on the intended application. The tract explains a wide range of functions, including military applications to gunnery and surveying. The details of the design allude to its military associations, for when folded it fits into a scabbard for carrying on a belt. The military surveyor or the gunner can thus draw his *radio latino* as readily as his sword.

Fig. 83 (bottom left)
The *radio latino*
applied to gunnery.
From catalogue no. 59.

Fig. 84 (below)
The *radio latino*
and its scabbard.
From catalogue no. 59.

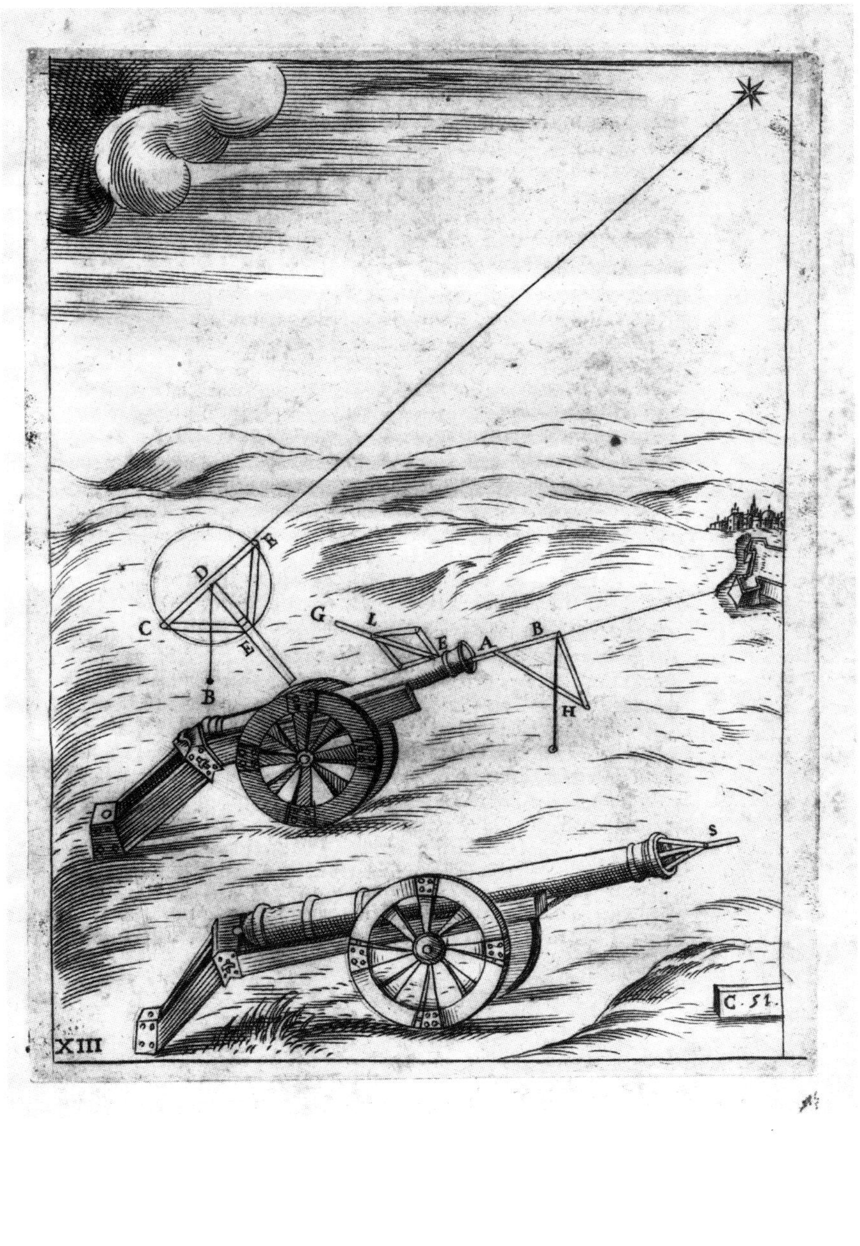

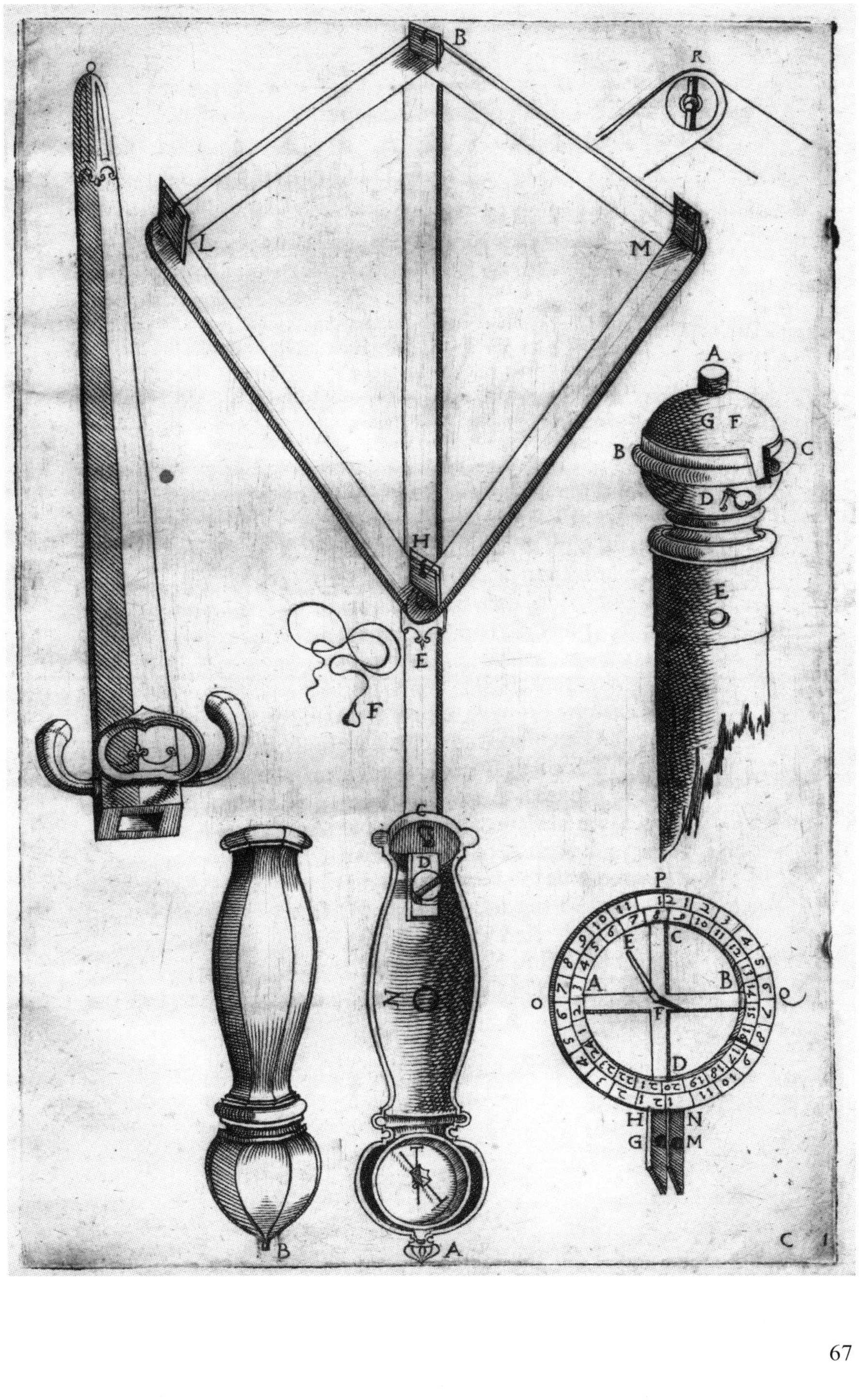

Fig. 85 (left)
The *radio latino* folded.
Catalogue no. 60.

Catalogue no. 60 is also illustrated, unfolded, on page 61.

Catalogue no. 61 is illustrated in colour on page 61.

60. Radio Latino

Signed: IO. MAR MANCINVS .F.
c. 1600

Among the scales on the pivoted arms of the instrument are those of the geometrical quadrant, marked 'VMBRA VERSA', 'VMB[RA] MED[IA]' and 'VMBRA RECTA'. The central arm, in addition to a degree scale, has positions marked for the number of sides of a polygon corresponding to the angle formed by two pivoted arms as arranged in the illustration (figure 82) to coincide with the outer angle of the corner of a fortification.

The handle is shaped, richly decorated and gilt. The silvered compass box incorporated into the handle is marked with the initials of the Italian names for the winds, and the reverse of this box has an engraving of Christ on the cross.

In its folded configuration (figure 85) the *radio latino* resembles a weapon and is carried in a scabbard (cf. figures 73 and 84).

Length (folded): 460 mm
Billmeir Collection
Inventory no. 37,525

61. Military Surveying Instrument

17th century

This instrument provides a link between more generalized tools of the surveyor and those specifically linked to fortification. A scale for laying out a polygonal structure on the ground or on paper must indicate, if not actual military use, at least an intention or acknowledgement of such a possibility.

Here the instrument is basically a simple theodolite (see catalogue no. 45), where a pivoted alidade moves over a circular degree scale. However, with the four fixed sights it also acts as a surveyor's cross or square, the reverse has a plumb-line, so that the instrument can be used as a level or for taking altitudes (a vertical orientation is permitted by the ball-and-socket joint), and the central compass for orientation is surrounded by the hour divisions of a sundial with a folding pin gnomon. In addition, an innermost scale, marked 'Polygon', sets out the angles made between the sides of three to twenty-sided regular polygons. There is a turned wooden case, lined with red velvet.

Diameter: 116 mm
Billmeir Collection
Inventory no. 36,730

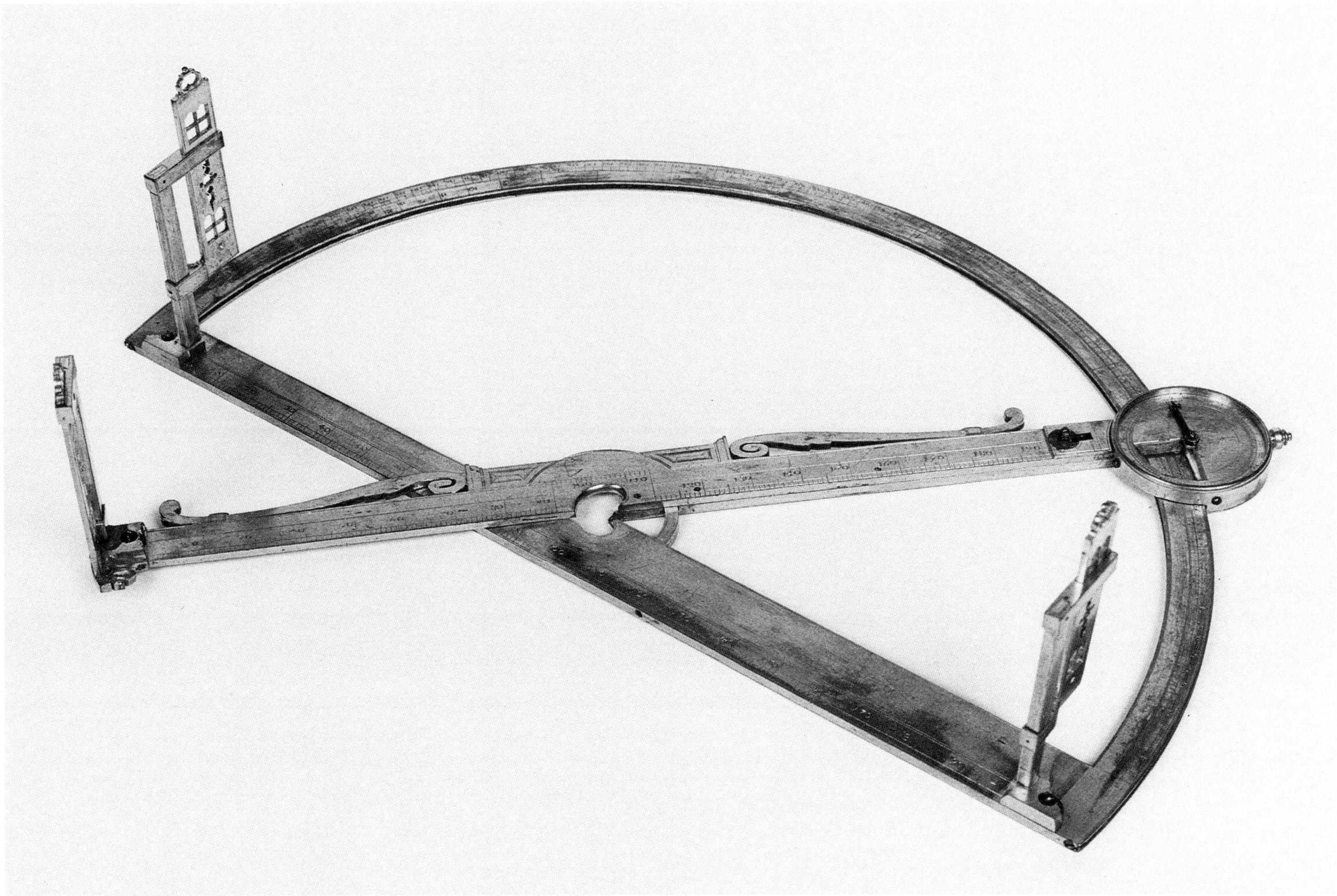

Fig. 86
Military graphometer and protractor.
Catalogue no. 62.

Catalogue no. 63 is illustrated in colour on page 64.

62. Military Graphometer and Protractor
17th century?

Another instrument which links general surveying with fortification is this graphometer with a scale for polygons. Both the diametric rule and the pivoted alidade have linear scales and pairs of complex sights, each of which moves on a short vertical scale (one sight is missing). The underneath of the instrument is flat and the alidade moves on an open ring rather than a central pivot, so that the centre of the semicircle is always exposed; the instrument can thus be used as a protractor.

The semicircular scale is divided to degrees and the subdivision of each degree into minutes is by a steel index arm, which is given an epicyclic motion over a circular scale divided 0–60 four times and rotates as the alidade is moved around the centre of the semicircle. The semicircle has a second set of divisions for laying out, in plan or in the field, the angles of regular polygons of between three and twenty-four sides.

Diameter: 300 mm
Billmeir Collection
Inventory no. 47,382

63. Surveying Instrument and Sundial
Signed: Paul Carre Fecit ÆParis 1652

This instrument is included to show how polygonal scales, originally justified as useful for laying out fortifications, can be conventionally included on instruments whose functions are becoming distant from those required for ordinary military surveying. The general arrangement here is similar to that of the simple theodolite (see catalogue no. 61), but the central compass is much larger and, having its own degree scale, can function as a circumferentor; it is now surmounted by a prominent equinoctial sundial, whose inclination is adjustable on a latitude scale. Shadow-square scales have been added to the scales on the top surface of the plate and the reverse has a plumb-line moving over a quadrant. The scale of 'Polygons' is of the same type and is in the same position as on catalogue no. 61, but here is for three to twelve-sided figures.

Dimensions: 200 x 200 mm
Billmeir Collection
Inventory no. 34,547

Fig. 87
Military protractor by Habermel.
Catalogue no. 65.

Catalogue no. 64 is illustrated in colour on page 64.

64. Surveying Instrument for Fortification
Signed: Ædescrolieres facie 1579

The three arms, each with a pair of sights, move on a single pivot and as they do so a cursor, connected to both outer arms, moves along the central arm, which is marked with a degree scale, indicating the angle between the outer arms, and with the number of sides of the regular polygon formed by repeating this angle. One of the outer arms bears the maker's signature, the other is engraved 'Pro declinatione muri', since these arms indicate the angle by which the walls of the fortification decline from each other. If the sights on the central arm are trained on the centre of the polygonal structure, those along the outer arms give the directions of the walls. A pin beneath is for mounting the instrument on a staff or tripod.

Length folded: 190 mm
Lewis Evans Collection
Inventory no. 36,261

65. Military Surveyor's Protractor
Signed: Erasmus Habermel fec:
Late 16th century

A beautifully executed, specialist protractor, by a maker whose instruments often have military associations. There are basically two scales on the brass plate – an outer scale of degrees subdivided by diagonals, and an inner set of scales for drawing polygons of from three to sixteen sides. To draw any of these regular polygons, the radial arm is set to the successive positions marked for the figure in question and the sides drawn with the tangential arm.

Diameter: 132 mm
Inventory no. 54,619

66. Military Architect's Rule
17th century

A German rule, shaped, pierced and decorated at one end, and bearing the monogram 'AF'. Both surfaces of the flat rectangle are crammed with tables and scales relevant to military architecture, recognising the different systems of three military engineers: 'Freytag', 'Vavban' and 'Klengel'. There is, for example, an extensive table of the sectional elements of a fortification, with the elements named and their proportions given for each system, as well as scales for setting out regular polygons. A.V. Simcock has suggested that 'AF' could be Adam Freitag, author of *Architectura militaris nova et aucta* (Leiden, 1631), whose system is included among those covered by this rule.

Length: 193 mm
Inventory no. 49,349

Figs 88 & 89
Both sides of a military
architect's rule.
Catalogue no. 66.

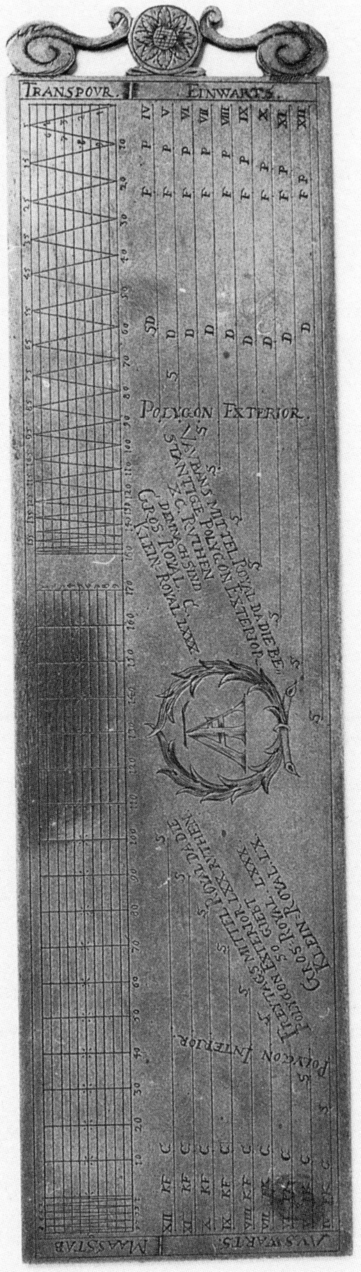

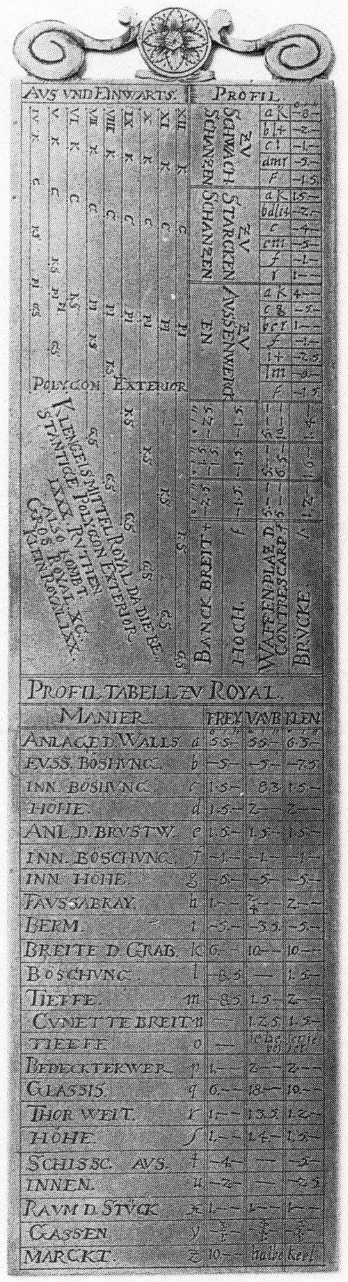

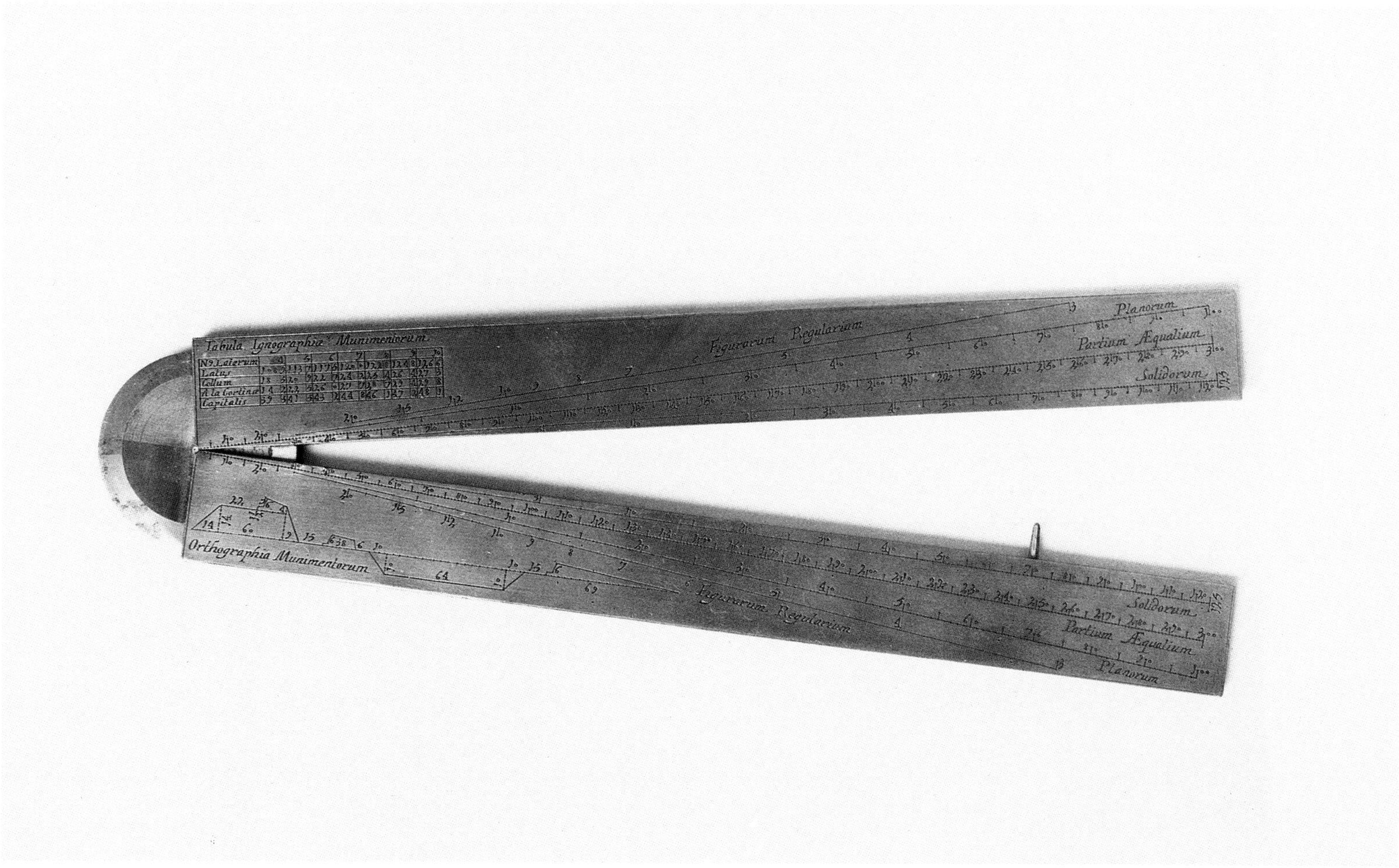

Fig. 90 (left)
Sector by Nicolaus Blondo, 1694.
Catalogue no. 67.

Fig. 91 (below)
Detail of the Blondo sector, showing the fortification table and diagram.
Catalogue no. 67.

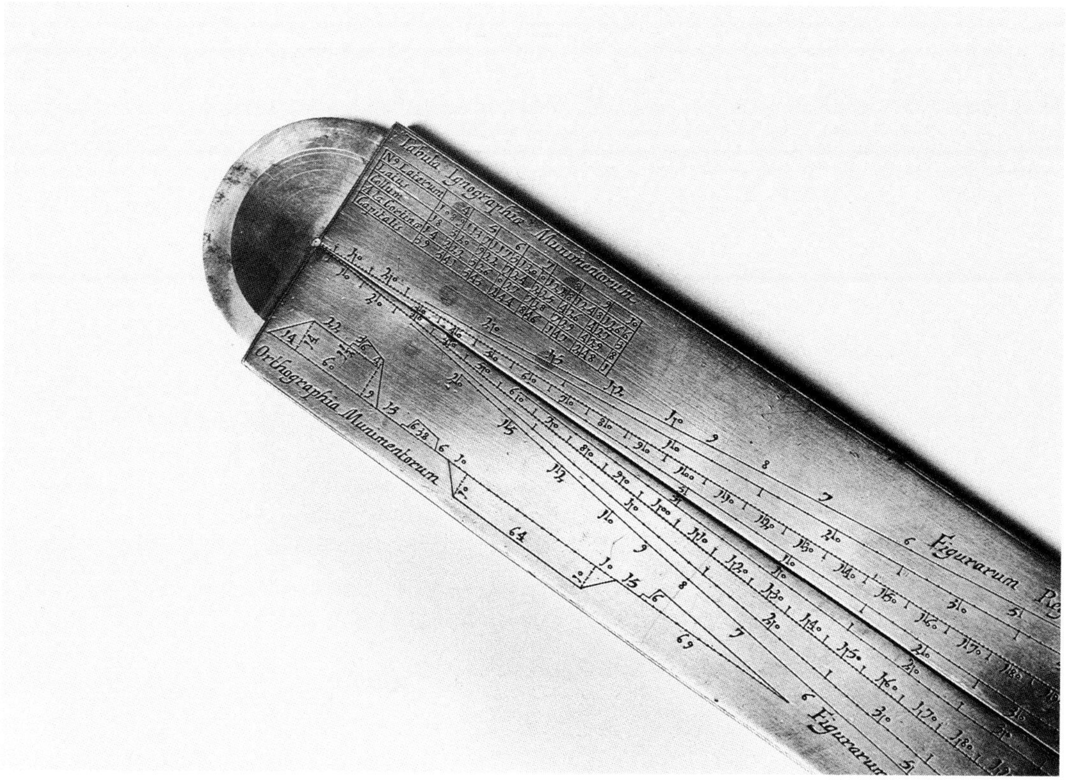

67. Sector

Signed: Nicolaus Blondó Fecit Neapoli A° 1694

This sector, in addition to a range of familiar scales for different calculations, which includes a pair of lines marked 'Figurarum Regularium' for setting out regular polygons with between three and twenty sides, has two additional features for the military surveyor. One, a 'Tabula Ignographiæ Munimentorum', gives the proportional dimensions of features of fortifications with between four and ten sides. The other, 'Orthographia Munimentorum', is a scaled section giving lengths, heights and depths according to a horizontal scale.

Length (folded): 231 mm
Inventory no. 49,132

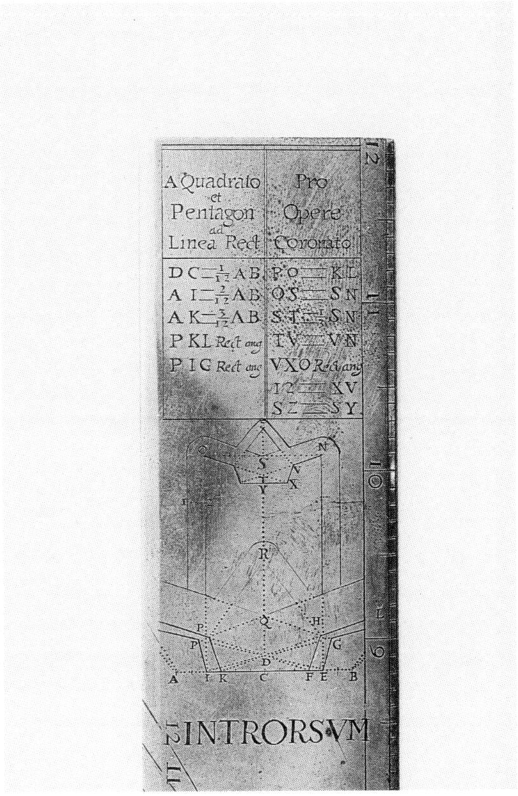

Figs 92 & 93 (left)
Details of both sides of a military architect's rule in brass.
Catalogue no. 68.

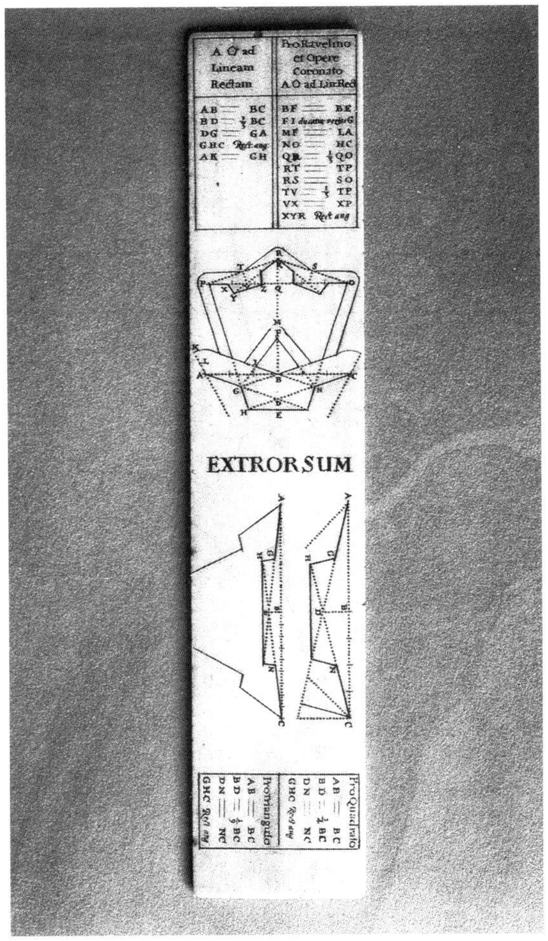

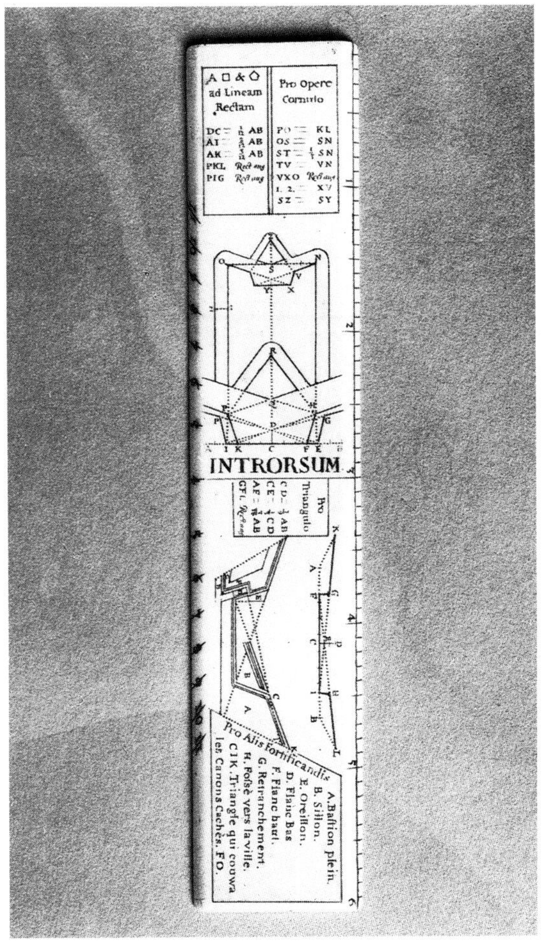

Figs 94 & 95 (below)
Both sides of a similar ivory rule.
Catalogue no. 69.

68. Military Architect's Rule
17th century

A brass rule, with a 12-inch scale on one bevelled edge, divided to 0.1, a protractor and a variety of fortification diagrams. The protractor is not for measuring angles in degrees, but is engraved with lines marked 12–4–12, with line 4 in the right-angle position. These lines are for setting out regular polygons with the indicated number of sides. Both sides of the rule have plans of fortifications with the geometry of their construction given in accompanying tables, one side marked 'INTRORSVM' (towards the inside), the other 'EXTRORSVM' (towards the outside).

Length: 304 mm
Inventory no. 45,283

69. Military Architect's Rule
c. 1700

An ivory rule, smaller than catalogue no. 68, but very similar in content. In fact the rule, protractor, fortification diagrams and tables are all equivalent, and similarly disposed between 'INTRORSUM' and 'EXTRORSUM', except that this rule has an additional diagram with an accompanying key marked: 'Pro Alis Fortificandis', giving the names of the various parts in French and, in one case, the function: 'Triangle qui couwa les Canons Cachè.'

Length: 152 mm
Inventory no. 33,529

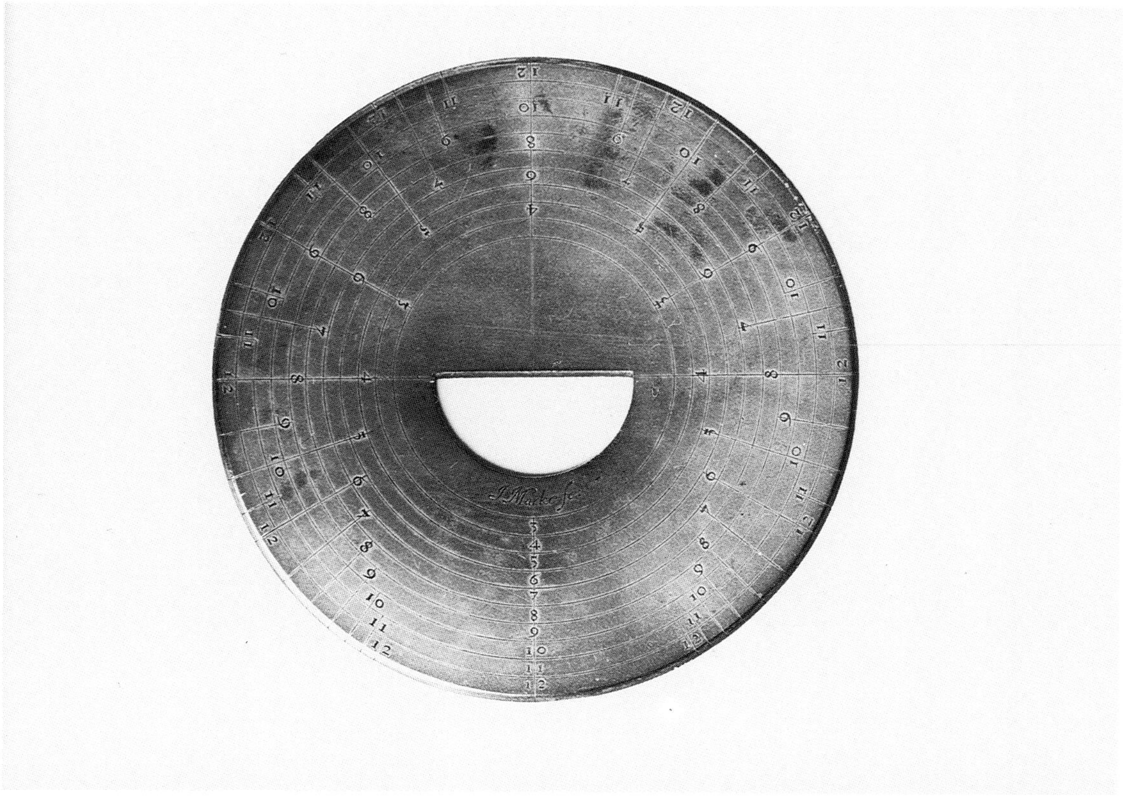

70. Military Protractor

Signed: I Marke fe:
c. 1670

A circular silver protractor with no other function than drawing polygons. The positions of lines from the centre to the corners of all regular polygons with between three and twelve sides are marked on ten concentric circles, each devoted to a single figure. There is no scale of degrees.

Diameter: 127 mm
Inventory no. 23,689

Fig. 96 (above)
English military protractor in silver.
Catalogue no. 70.

Fig. 97 (below)
Protractor for fortifications by John Rowley.
Catalogue no. 71.

71. Protractor for Internal and External Angles

Signed: I. Rowley fecit
c. 1700

The semicircular scale is divided to half-degrees and numbered by 10 in both directions, i.e. 0–180 and 180–0. The set of jointed parallel rules relate the internal angle, external angle, and angle subtended at the centre for the walls of a regular polygonal fortification. This form of instrument was known in England as a 'parallelogram protractor'. Another instrument maker, Thomas Tuttell described it in his *Description and Explanation of Mathematical Instruments* (1701) and explained that the semicircular scale indicates 'the quantity of any inward or outward angle'.

Diameter: 174 mm
Orrery Collection (Christ Church)
Inventory no. 32,319

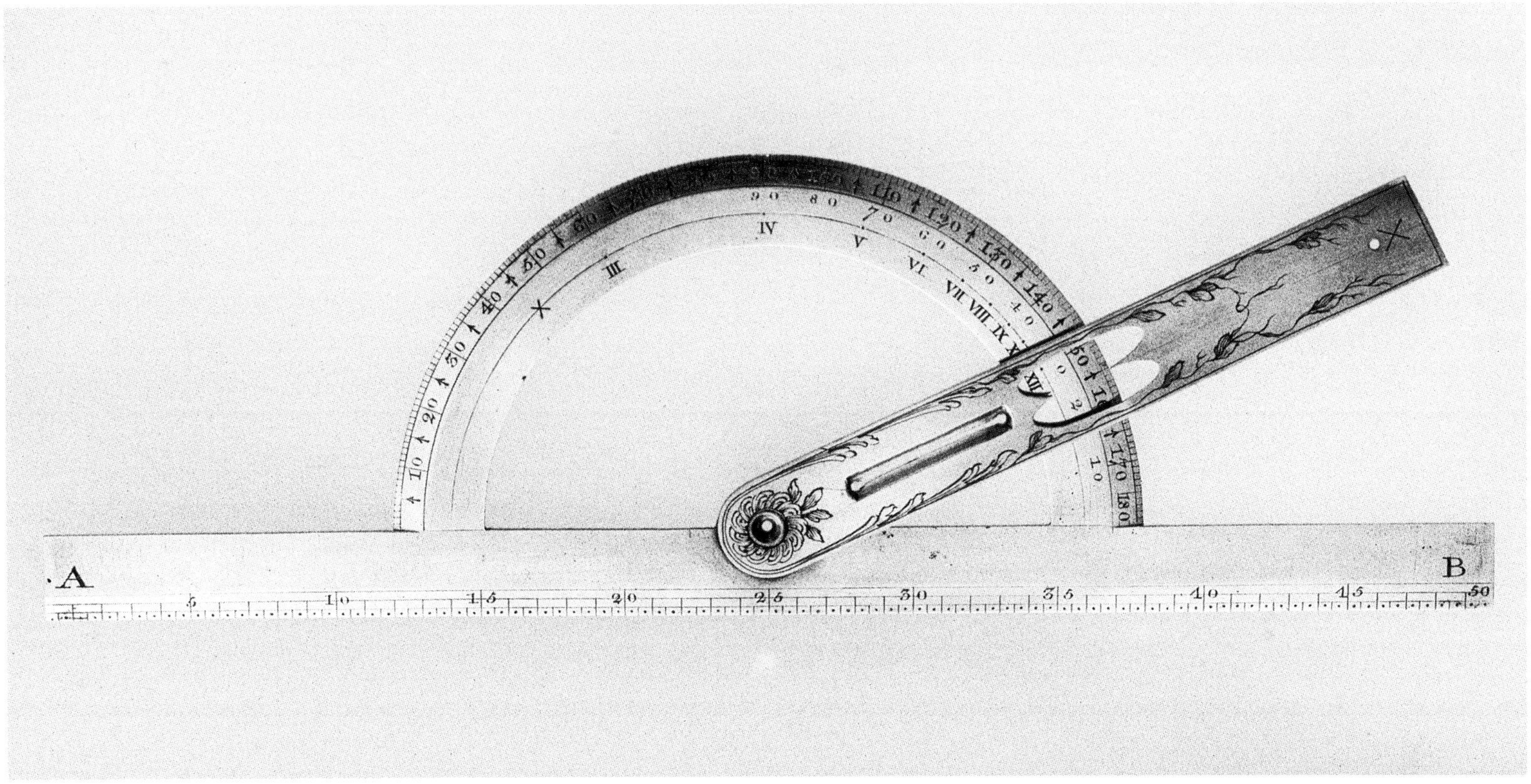

Fig. 98 (right)
English architect's protractor in silver, with polygonal scale.
Catalogue no. 73.

Fig. 99 (below)
German military protractor, 1712.
Catalogue no. 72.

72. Military Protractor
Signed: 1712 Johan Martin Vnseld. Vlm fecit.

A German protractor with a semicircle graduated in degrees and marked with positions for regular polygons of between four and twenty sides. Apart from '4', at the centre (90 degrees), representing the square, each polygon has two marks distributed symmetrically about the right-angle position, so that the marks run from 20 to 4 to 20. These positions give both the interior angles between the walls and the complementary angles subtended by the walls at the centre of the polygon.

Diameter: 100 mm
Billmeir Collection
Inventory no. 32,196

73. Architect's Protractor
Signed: Joshua Kirby Inv^{t}. Octr. 5. 1765

This small silver protractor, with positions for three to twelve-sided polygons alongside the usual degree scale, illustrates how such polygonal scales continue to appear on architectural instruments into the later eighteenth century.

Length: 152 mm
Inventory no. 23,312

Fig. 100
An elevated fortress.
From catalogue no. 74.

74. Joseph Furtenbach, *Mannhaffter Kunst-Spiegel* (Augsburg, 1663).

A German compendium of the mathematical arts, with a strong emphasis on practical and mechanical matters. Furtenbach published texts on many individual topics, especially on various aspects of architecture, from civil to military to 'universal architecture'. In this larger work he referred back constantly to his prior publications, treating his text as a summation and synthesis of his earlier concerns. The arts of war figure prominently, through the discussion not just of fortresses on land but also of the fortification of islands. Furtenbach also considers naval architecture, gunnery and 'fireworks', in addition to many other mathematical arts such as astronomy, geography, perspective and navigation.

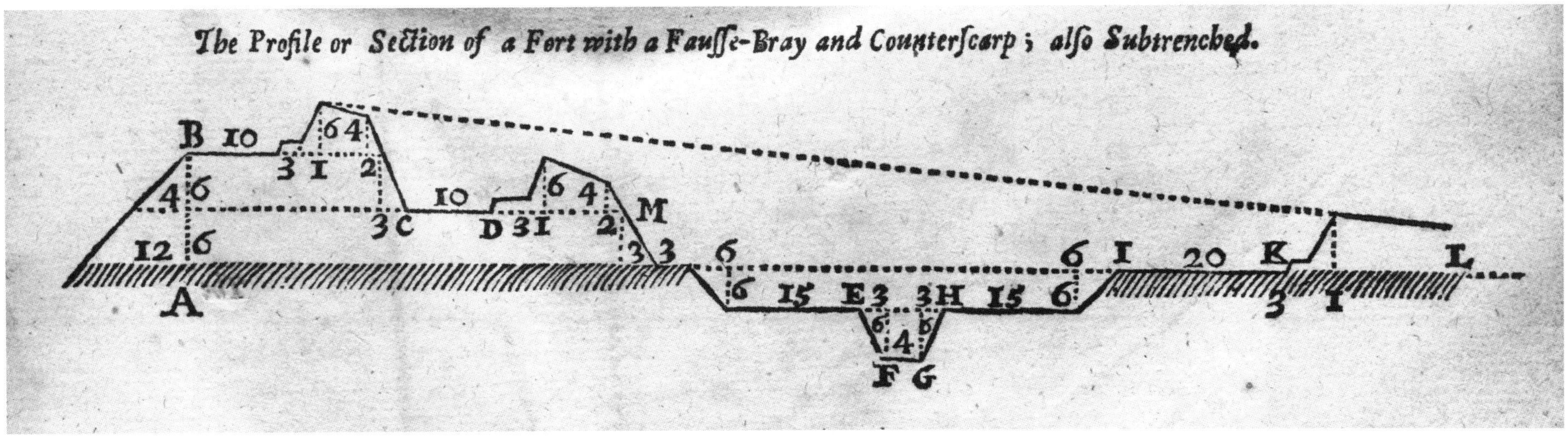

Fig. 101 (above)
Section of a fortress with its elaborate ditch and outworks.
From catalogue no. 75.

Fig. 102 (below)
A system of fortification and its protagonist.
From catalogue no. 76.

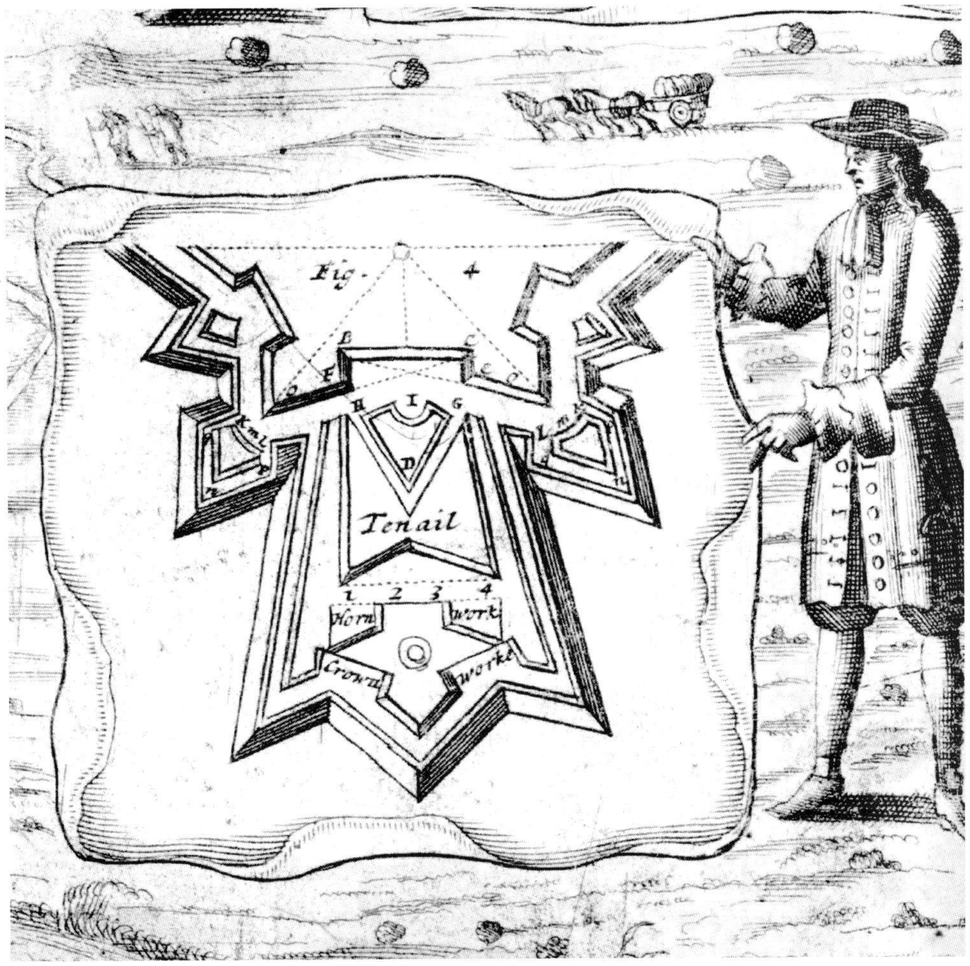

75. Philip Staynred, *A Compendium of Fortification: both Geometrically and Instrumentally by a Scale* (London, 1683).

It was typical of the mathematical practitioners, whose numbers were growing significantly in early-modern England, to seek to design instruments to help their clients avoid too much calculation. Here Staynred, 'Professor and Teacher of the Mathematicks in the City of Bristol', explains the geometrical construction of fortifications, the use of tables he provides for calculating their elements, and the construction of an instrument, 'The Scale of Fortification', from these tables. His short tract was printed in London by Mary Clark, and introduced by Samuel Sturmy into his *Mariners Magazine* (catalogue no. 41). The title-page carries a verse from the Book of Jeremiah: 'The Lord is my strength, and my fortress, and my refuge in the day of affliction'.

76. Richard Blome, *The Gentlemans Recreation* (London, 1686).

Blome included fortification in his vast compendium of the arts and sciences recommended to the English gentleman, and he had the relevant section written by George Leslie. A variety of alternative methods and systems are presented, indicating different authorities and traditions of practice within the developing domain of fortification.

77. Johann Christoph Sturm, *Mathesis compendiaria* (Coburg, 1714).

It was clearly considered important for a gentleman to know the proper terminology of fortification. In this mathematical compendium components are carefully named, and equivalents given in Latin, French and German. Such terminological emphasis adds to the impression given by the geometrical formalism, to suggest that increasingly fortification could be presented as much as a polite science as a practical one.

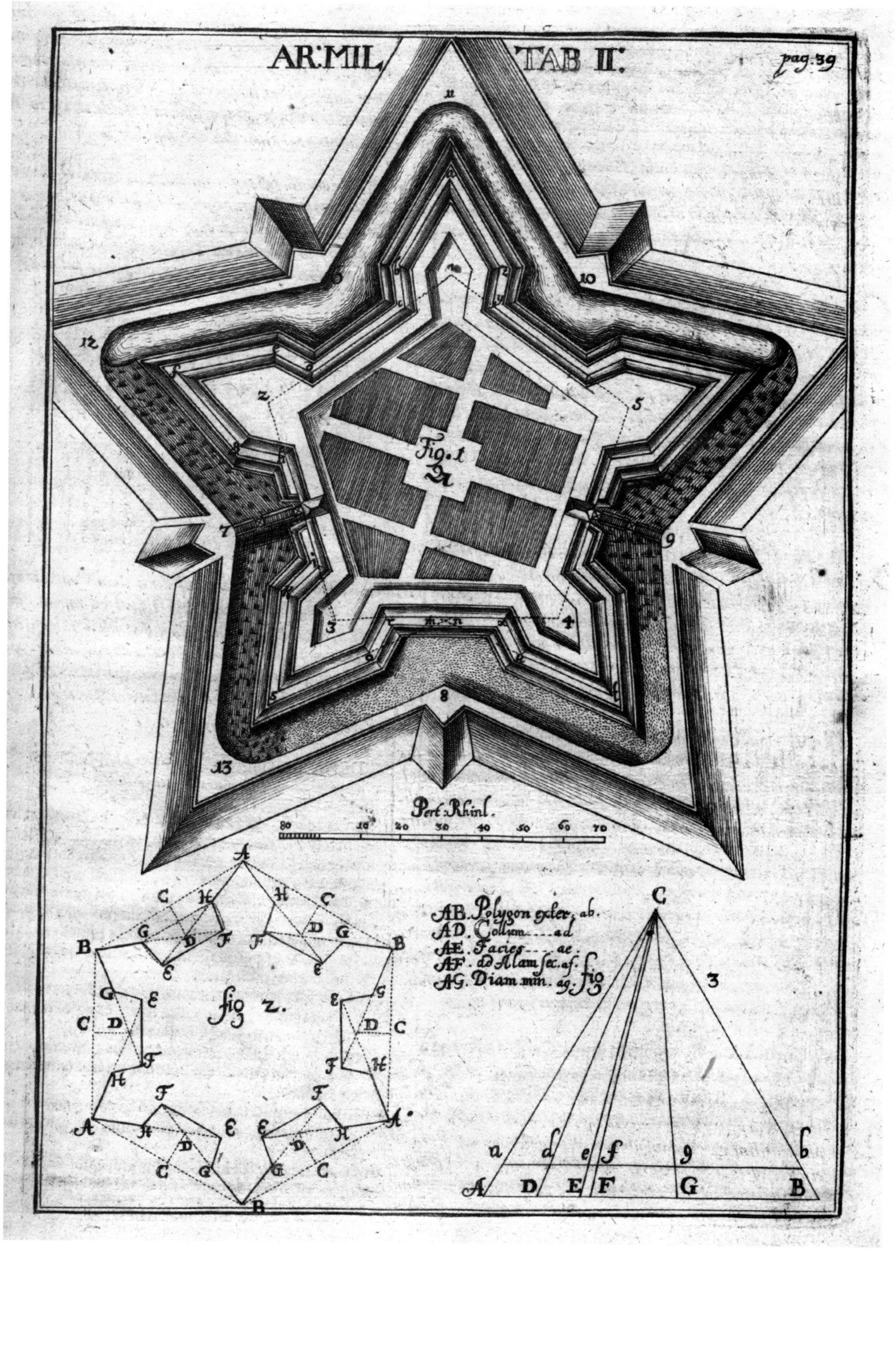

78. William Oughtred, *The Circles of Proportion and the Horizontal Instrument* (first edition, 1632; this edition, London, 1633).

The ordering of soldiers in regular formations was a frequent topic of mathematical and military discussion. It was regularly included in expositions of the use of particular instruments, for example the 'military proteus' (catalogue no. 12). It also came within the bounds of William Oughtred's account of the manifold calculating uses of his 'circles of proportion'. The instrument consisted of a series of concentric logarithmic scales operated with rotating indices, and was in effect a form of slide rule. Oughtred provided some numerical examples on the calculation of troop formations amongst material that dealt with arithmetic, geometrical problems of plane and solid measurement, gauging, astronomy, and trigonometry. He pretended to no great military expertise but included his treatment principally as an exercise in the use of his instrument.

This copy of Oughtred's text has, as well as the book's title, an additional stamped inscription on the spine: 'CORRECT BY HIS OWN HAND'. The annotations do indeed appear to be by the author, since they record many emendations of which some (but not all) were featured in later editions of the book.

Fig. 103 (left)
Fortification based on a pentagonal plan.
From catalogue no. 77.

Fig. 104 (below)
Oughtred's corrected text on troop formations.
From catalogue no. 78.

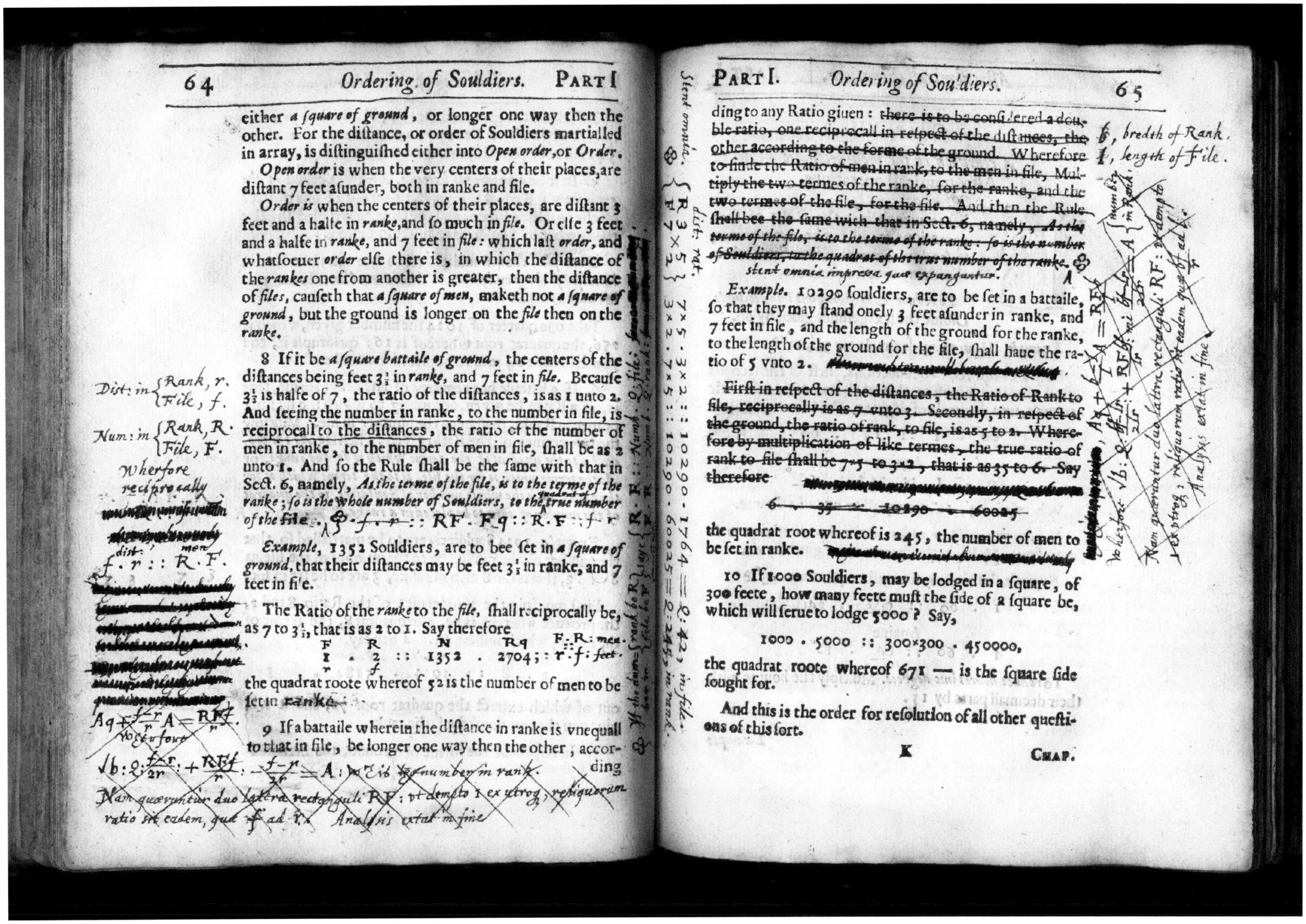
64 *Ordering of Souldiers.* PART I

either *a square of ground*, or longer one way then the other. For the distance, or order of Souldiers martialled in array, is distinguished either into *Open order*, or *Order*.

Open order is when the very centers of their places, are distant 7 feet asunder, both in ranke and file.

Order is when the centers of their places, are distant 3 feet and a halfe in *ranke*, and so much in *file*. Or else 3 feet and a halfe in *ranke*, and 7 feet in *file*: which last *order*, and whatsoeuer *order* else there is, in which the distance of the *rankes* one from another is greater, then the distance of *files*, causeth that *a square of men*, maketh not *a square of ground*, but the ground is longer on the *file* then on the *ranke*.

8 If it be *a square battaile of ground*, the centers of the distances being feet 3½ in *ranke*, and 7 feet in *file*. Because 3½ is halfe of 7, the ratio of the distances, is as 1 unto 2. And seeing the number in ranke, to the number in file, is reciprocall to the distances, the ratio of the number of men in ranke, to the number of men in file, shall be as 2 unto 1. And so the Rule shall be the same with that in Sect. 6, namely, *As the terme of the file, is to the terme of the ranke; so is the whole number of Souldiers, to the true number of the file.*

Example, 1352 Souldiers, are to bee set in *a square of ground*, that their distances may be feet 3½ in ranke, and 7 feet in file.

The Ratio of the *ranke* to the *file*, shall reciprocally be, as 7 to 3½, that is as 2 to 1. Say therefore

F . R :: N . Rq

1 . 2 :: 1352 . 2704;

the quadrat roote whereof 52 is the number of men to be set in ranke.

9 If a battaile wherein the distance in ranke is vnequall to that in file, be longer one way then the other, accor- ding

PART I. *Ordering of Souldiers.* 65

ding to any Ratio giuen: there is to be considered a double ratio, one reciprocall in respect of the distances, the other according to the forme of the ground. Wherefore to finde the Ratio of men in ranke, to the men in file, Multiply the two termes of the ranke, for the ranke, and the two termes of the file, for the file. And then the Rule shall bee the same with that in Sect. 6, namely, *As the terme of the file, is to the terme of the ranke: so is the number of Souldiers, to the quadrat of the true number of the ranke.*

Example. 10290 souldiers, are to be set in a battaile, so that they may stand onely 3 feet asunder in ranke, and 7 feet in file, and the length of the ground for the ranke, to the length of the ground for the file, shall haue the ratio of 5 vnto 2. [redacted]

First in respect of the distances, the Ratio of Ranke to file, reciprocally is as 7 vnto 3. Secondly, in respect of the ground, the ratio of ranke, to file, is as 5 to 2. Wherefore by multiplication of like termes, the true ratio of ranke to file shall be 7·5 to 3·2, that is as 35 to 6. Say therefore [redacted]

6 . 35 :: 10290 . 60025

the quadrat root whereof is 245, the number of men to be set in ranke. [redacted]

10 If 1000 Souldiers, may be lodged in a square, of 300 feete, how many feete must the side of a square be, which will serue to lodge 5000? Say,

1000 . 5000 :: 300×300 . 450000,

the quadrat roote whereof 671 — is the square side sought for.

And this is the order for resolution of all other questions of this sort.

K CHAP.

Fig. 105
Circles of proportion
by Elias Allen.
Catalogue no. 79.

79. Circles of Proportion
Signed: Elias Allen fecit
c. 1635

As well as describing his 'circles of proportion', William Oughtred's text specified another instrument to be engraved on the reverse of the plate: the 'horizontal instrument', which is based on a projection of the celestial sphere. This example by Elias Allen, the leading English instrument maker of his day and a close collaborator of Oughtred, has both the 'circles of proportion' and the 'horizontal instrument'. The circles are in the revised form that was published in 1633 (catalogue no. 78), with the inclusion of two extra circles described in an additional section on navigation. Rather than a nocturnal as shown in the printed version of the instrument, there is a planispheric projection of the celestial sphere of the Rojas type inside the innermost circle.

This example of the instrument came almost directly to Oxford after being made: it was given to St John's College by George Barkham and an inscription records the date as 1635. Below the inscription Barkham's arms are prominently engraved. Not only does this provenance demonstrate the academic context of this particular example, but its large size strongly suggests that it would have served primarily for study and practice rather than for more active exertions.

Radius: 230 mm
St John's College Collection
Inventory no. 40,847

Fig. 106 (above)
Military counters.
Catalogue no. 80.

80. Military Counters

c. 1700

When the 18th-century English instrument maker Thomas Wright catalogued the collection to which this set of brass plates belongs, he was evidently at a loss as to their function. They clearly did not belong to the contemporary instrument-making repertoire and Wright could suggest only that they bore some relation to gunnery. A more plausible explanation is that they were used to display the disposition of troops.

There are five full-length and four half-length plates, with the shorter pieces marked 'Grenadiers' and numbered from 1 to 4. Four of the longer plates each represent a 'Division' and are numbered in sequence from the first to the fourth, each subdivided into three further sections. All also have 'Angle' marked at one end of the plate in such a way that they are most readily assembled into a square. The final plate, with a table of numbers, perhaps displays an arrangement of these or similar plates.

The plates are stored in a simple wooden case.

Dimensions (full-length plates): 113 x 18 mm
Orrery Collection (Christ Church)
Inventory no. 50,368

Fig. 107
Replica of a
Galilean telescope.
Catalogue no. 81.

81. Telescope
Modern replica

The exhibition ends not with an instrument from the repertoire of practical mathematics but with a telescope. The invention of the telescope inaugurated a novel class of scientific instruments. These new optical instruments, which soon included the microscope as well as the telescope, were distinct from the traditional instruments of practical mathematics through their use of lenses and mirrors. They also quickly came to be manufactured by a distinct group of optical instrument makers. Despite such differences, optical instruments such as the telescope were, like mathematical instruments, intimately bound up with military preoccupations.

The first public announcement of the telescope was a printed newsletter describing the visit of an embassy of the King of Siam to Prince Maurice of the Netherlands in October 1608 – an unlikely setting for an instrument that is now one of the classic icons of science. But it was as a military instrument that the new invention was initially perceived and the newsletter was quite explicit about the purpose of the device. 'The said glasses are very useful in sieges and similar occasions, for from a mile and more away one can detect all things as distinctly as if they were very close to us.'

The publication of the newsletter destroyed efforts to keep the existence and construction of the telescope a secret and reports of the marvellous novelty were transmitted with incredulity and excitement. Within a few months copies of the telescope were being sold in several European cities.

On hearing news of the invention, Galileo set about constructing and improving his own version. After presenting his telescope to the Venetian Senate, Galileo was honoured with a promise that his salary be doubled and his appointment at the University of Padua confirmed for life. These rich rewards were offered before he had made his astronomical discoveries; Venice was recognising the strategic rather than scientific importance of Galileo's improvements.

Galileo already well knew the value of instruments within the realm of patronage. His sector or 'geometric and military compass' was a prime example. Devised for military uses in 1597 and afterwards revised, examples were offered to both the Medici and the Gonzaga courts. It also formed the basis of private courses taught by Galileo in Padua in the first years of the 17th century, though his military teaching included fortification as well. The sector was an important device for practical mathematics and several military versions appear elsewhere in the exhibition. But it could not match the telescope for spectacle and the wide-ranging implications of the observations made with its help.

Perhaps prophetically, the original 1608 report mentioned that, in addition to its evident military worth, the new device could also be turned on the heavens. Months before Galileo had even heard of the invention it was already being reported that 'even the stars which ordinarily are invisible to our sight and our eyes, because of their smallness and the weakness of our sight, can be seen by means of this instrument.' The results of observing the celestial appearances not only propelled Galileo into a new position as Mathematician and Philosopher at the Medici court but redirected contemporary debate in astronomy and natural philosophy. From a military device whose dissemination states sought to control, the telescope became a more subtly destructive instrument, used to provide vital new evidence for the Copernican cosmology. For a figure such as Galileo, it was only a short step from the geometry of war to the new science of the 17th century.

This instrument is a copy of one of Galileo's telescopes preserved in Florence. The drawtubes are of wood covered with decorated vellum and leather and the magnification is about x20.

Length (closed): 923 mm
Inventory no. 37,599

Bibliography

John H. Appleby 'A New Perspective on John Rowley, Virtuoso Master of Mechanics and Hydraulic Engineer', *Annals of Science*, 53 (1996), pp. 1–27.

Mario Biagioli 'Galileo's System of Patronage', *History of Science*, 28 (1990), pp. 1–62.

D. J. Bryden 'The Medal as a Technical *aide memoire*: an Example from Restoration England', *The Medal* (forthcoming).

Franz Adrian Dreier *Winkelmessinstrumente* (Berlin, 1979).

Wolfgang Eckhardt 'Erasmus und Josua Habermel – kunstgeschichtliche Anmerkungen zu den Werken der beiden Instrumentenmacher', *Jahrbuch der Hamburger Kunstsammlungen*, 22 (1977), pp. 13–74.

Galileo Galilei *Operations of the Geometric and Military Compass*, trans. Stillman Drake (Washington, 1978).

J. R. Hale *Renaissance Fortification: Art or Engineering?* (London, 1977).

A. R. Hall *Ballistics in the Seventeenth Century* (Cambridge, 1952).

I. Hogg *The History of Fortification* (London, 1981).

Q. Hughes *Military Architecture* (2nd ed., London, 1991).

John R. Millburn 'The Office of Ordnance and the Instrument-Making Trade in the Mid-Eighteenth Century', *Annals of Science*, 45 (1988), pp. 221–93.

John R. Millburn 'Gunnery Instruments and their Makers in the Eighteenth Century', *Journal of the Ordnance Society*, 7 (1995), pp. 27–44.

Mara Miniati (ed.) *L'età di Galileo: Il secolo d'oro della scienza in Toscana* (Florence, 1987).

Alan Q. Morton & Jane A. Wess *Public and Private Science: The King George III Collection* (Oxford, 1993).

I. Newton *A Treatise of the System of the World* (London, 1728).

R. Norton *The Gunner* (London, 1628).

Geoffrey Parker *The Military Revolution: Military Innovation and the Rise of the West, 1500–1800* (Cambridge, 1988).

Simon Pepper & Nicholas Adams *Firearms and Fortifications: Military Architecture and Siege Warfare in Sixteenth-Century Siena* (Chicago, 1986).

Charles Singer & Trevor I. Williams (eds) *A History of Technology* (7 vols, Oxford, 1954–84), vol. 3, *From the Renaissance to the Industrial Revolution* (Oxford, 1957).

Brett D. Steele 'Muskets and Pendulums: Benjamin Robins, Leonhard Euler, and the Ballistics Revolution', *Technology and Culture*, 35 (1994), pp. 348–82.

E. G. R. Taylor *The Mathematical Practitioners of Tudor & Stuart England* (Cambridge, 1954).

Albert van Helden 'The Invention of the Telescope', *Transactions of the American Philosophical Society*, 67 part 4 (1977).

Hélène Vérin *La Gloire des ingénieurs: l'intelligence technique du XVIe au XVIIIe siècle* (Paris, 1993).

Herbert Wunderlich *Kursächsische Feldmesskunst, artillerische Richtverfahren und Ballistik im 16. und 17. Jahrhundert* (Berlin, 1977).

Ernst Zinner *Deutsche und niederländische astronomische Instrumente des 11. bis 18. Jahrhunderts* (2nd ed., Munich, 1967).